Is Reality a Simulation? An Anthology

Edited by

Antonin Tuynman Ph.D.

I0461980

Published by Antonin Tuynman
Rijswijk
Netherlands

Cover by Ramon J. Tuynman, © Antonin Tuynman 2018

Contents

Preface

The origins of this book go back to 2010 when I started writing a blog called "Technovedanta" about the possibility that the internet might one day awaken as a conscious, omnipresent and omniscient conscious entity. I had been inspired by reading several books in the field of A.I. and started to bring my ideas to the attention of a public on a futurism site called "KurzweilAI". On this site there was a forum, where like-minded people discussed about the upcoming technological singularity (the point in history, probably triggered by a runaway of superintelligent AI in conjunction with Transhumanism and nanotechnology, after which the changes due to technological growth to our human civilization become unpredictable).

It is here that I met people with a great imagination such as Matt Swayne, Tim Gross and Sean Byrne, who were pioneers peeking beyond the veil of the speculative post-singularity. A topic discussed ever so often regarded the question whether we were already living in a computer simulation - of course with frequent references to the popular 1999 movie "the Matrix". Tim Gross published his brilliant "Ouroboric Simulist Cosmology" on this forum, but this year the forum was taken down integrally thereby ruthlessly destroying all the magnificent contributions from the Singularitarians. Fortunately, this book will give you an opportunity to taste the saved nectar of their teachings.

As a consequence of these interactions I wrote various essays on my blog, which I decided to bundle in a book which I published in 2012 under the title: "Technovedanta, Internet Architecture of a Quasiconscious Webmind, A Panpsychic Theory of Everything". The book proposed how the internet could be enhanced to become a webmind by imparting an architecture to it, which I based on a stratification of the mind derived from Indian philosophy, in particular from Vedanta.

One of the books that had inspired me to write Technovedanta, was the astonishing book "Technomage" about the interface between technology and occultism, written by Dirk Bruere, who was also a contributor to the KurzweilAI forum.

After reading my book, my friend Matt Swayne from the KurzweilAI forum, expressed the wish that he'd love to see a sequel, preferably written jointly by me and Tim Gross.

In 2017 I published a sequel to this book called "Technovedanta 2.0, Transcendental Metaphysics of Pancomputational Panpsychism". In both Technovedanta books I already suggest, that we might live in a computer simulation, although probably not run on a von Neumann computer like architecture.

Matt Swayne made me aware, that some of my posts had been picked up by Alex Vikoulov, who referred to them on his site Ecstadelic Media. Alex has been writing various mind-boggling essays on similar subject matter, which he plans to publish in the near future in the form of a book. A frequent topic Alex addresses is of course the simulation hypothesis.

In the same year, my book "Is Intelligence an Algorithm?", which addresses some of the questions of my first book in a more accessible format, was accepted for publication by iff-books.

As a consequence I was contacted by Knujon Mapson, another author at iff-books who had compiled an Anthology on Pandeism. We agreed to review each other's books. It was Knujon's excellent anthology, which gave me the idea to compile the present anthology on the simulation hypothesis.

In June 2017 Bernardo Kastrup, a well-known idealist philosopher and author for iff-books published a guest essay written by me titled: "Is Panpsychism Irreconcilable with Idealism". This essay was discussed and reviewed in Bernardo's forum "Metaphysical Speculations" where I met the idealist philosopher Dante Rosati.

In November 2017 I launched a call on the internet for authors to contribute to the present anthology "Is Reality a Simulation". The assignment was to write an essay, which deepened the notions referred to in this question in a manner well beyond Bostrom's traditional "simulation argument". I addressed about twenty different authors.

Of course I invited my friends Matt Swayne, Tim Gross and Sean Byrne from the KurzweilAI forum, as well as Knuje and Alex, who fortunately all answered to my summoning call from the hellish depths of the Cyberbardo.

That year I had also read the fascinating book "The science of Consciousness" by Eva Deli and I invited her to contribute too, which invitation she fortunately accepted.

Sean asked me whether his friend Donald King, who had developed his own unorthodox and provocative philosophy, could contribute too and I'm glad I consented to this request.

In a second call I was able to drag Dirk Bruere and Dante Rosati into the rabbit hole of writing essays about the simulation hypothesis, so that in the beginning of May 2018 I could start compiling an anthology on this topic. Because my calls had not been able to rally all the authors I had invited and the contributions had not provided me with the number of pages I was aiming for, I have had to enrich this book with a couple of extra chapters written by me, which I had previously published as essays on my blog.

Note that some of the essays may be quite unconventional and even controversial. Some of them express diametrically opposing perspectives. Their inclusion in this anthology does not mean that I, as editor and co-author necessarily agree with these ideas. Nor is it an admission of their "scientific" correctness. If this book aims to achieve anything, it is to widen your horizon and to contemplate challenging alternatives, rather than to lock you up in the ruling paradigms of science.

It is a great honor for me to be able to present you with this Great Work of a delicate selection of great minds. Whereas the topic of the simulation hypothesis has been addressed from various angles by various thinkers over the ages, I am convinced that the present volume can still be a blood-curdling and nerve-racking adventure into the outer confines of the human imagination, as the authors of this volume have kept their promise to provide me with an unorthodox and utterly original account of their perspective on this mind-boggling subject-matter.

The Quest for Reality

Did your dark night of the soul ever make you doubt the reality of your existence? Do you wonder whether you are living in a dream or computer simulation? Are you haunted by the perspective that you're already dead and wander through the infinite dimensions of Hell or the Cyberbardo? Does it really matter at all if one of these questions is answered affirmatively?

Then you're in good company. Join us on a psychedelic rollercoaster through the rabbit hole. Fasten your seatbelts. Your belief systems are about to be questioned, challenged and perhaps overthrown. This Anthology with contributions from Technoshamans, Physicalist scientists, Pantheists, Pandeists and Panpsychists will rock your mental foundations, haunt your convictions and put you through the epistemological wringer.

This choicest selection will sharpen your mind to find truths hidden in the plain sight of a tower of turtles, patterns in a grid of chaos and clarity in a forest of apparent randomness. From Gross' Ouroboric Simulism to the Other of Swayne; Dive into Rosati's lucubration from Deli's Fractal of Consciousness to Mapson's Pandeistic Analogue Simulation; from the Vikoulovian Apotheosis via Byrne's Panpsychic Musings to the Tuynman Omega Constant. Escape with Bruere's scenario's from King's Parasitism and Perceptual valuation.

Welcome to a dazzling orgy of the post-singularity conceptualization of Simulation Theory. Welcome to the kaleidoscopic variegation of the perplexing pictorial perspectives that dwarf Bostrom's argument into oblivion.

Chapter 1

An introduction into the quest for Reality

Antonin Tuynman studied Chemistry at the University of Amsterdam, achieving both an MSc and a PhD, and worked as a postdoc researcher at the "Université René Descartes Paris V" in Paris. Since 2000, Tuynman has worked as a patent examiner at the European Patent Office (EPO) in the field of clinical diagnostics. He has vast experience in meditation and yoga, and a strong interest in Hinduism and Buddhism. He also has strong affinity for futurism and the Singularity theory of Kurzweil. In his books, Tuynman proposes Artificial Intelligence concepts which may lead to the emergence of internet as a conscious entity using stratifications from Vedic scriptures.

You can't say "Grab 'em by the pussy" and then get elected president. No way. This must be a glitch in the Matrix. Something has gone haywire. The controllers or Gods or whoever they are, have either lost it or are pulling our legs. This is a joke of a prankster, intervening with the plans of the architect.

Such arguments were circulating on the internet just after Trump got elected president of the United States of America. It shows how deeply the concept that we might be living in some kind of simulated virtual reality got rooted in less than two decades since the release of the epic science fiction film the Matrix in 1999.

Although that movie may have been the point at which this concept really became popular among the masses, it may have been around since the birth of culture and language in human society. After all, if you believe in a God or multiple Gods who created this world, you are somehow admitting that this world is just a product of imagination.

In certain branches of Hinduism it is believed that Vishnu dreams up a plethora of worlds. What we call reality is considered to be "Maya", an illusion. The Taoist Zhuangzi wondered in the fourth century B.C. whether he was a man dreaming he was a butterfly or a butterfly dreaming he was a man and around the same time, Plato described prisoners who took the shadows on the wall of the cave in which they were chained for reality.

So the idea that reality might not be so "real" after all is not a new invention of the computer era. Of course Nick Bostrom's "Simulation Argument", which I'll discuss in a later chapter in more detail and the technological singularity movement of "Technopapes" such as Vernor Vinge, Ray Kurzweil and Elon Musk have boosted the modern "reality-is-a-computer-simulation" version of this notion, but as you will discover in the chapter on the history of Simulation Theories, this is merely a continuation of man's age-old deep philosophical quest to understand himself and the world around him.

But even if it were true that we are living in a simulation, would that really change something about the way we conduct our lives? And what can yet another book on this topic bring you? Is the answer to such a question not forever elusive and outside the grasp of what can be known at all? Does reality have in-built mechanisms to veil its underlying nature, so that an unveiling or "Apokalypsis" as the Greek used to call it, is an impossibility? Or can we actually find some strong indicators and pointers that at least ground our strong suspicions about the simulation hypothesis or even find arguments that convince us? Can we extrapolate the implications of this realization, if it turns out to be the truth?

If you are curious, hungry or yearning to see if these questions can be answered, you are probably the right audience for this book. If not, don't waste your money and time and put the book back on the shelf now. You can still take the blue pill. Do it!

Now that you have decided to continue to read, you find yourself in the comforting company of those who have asked the same questions. Authors, philosophers, scientists and normal curious people who would like to get to the bottom of the questions: What is Reality? Am I real? Who or what am I and what is this world around me? You are a truth seeker. Glad to have you on board, so are we.

I used "we" to indicate the authors of this anthology. Rather than giving you a one-sided perspective of what I, Antonin Tuynman (the editor of this book and author of a few chapters herein) have thought of this matter, I decided to provide you with a diverse overview of views of a number of thinkers ranging from down-to-Earth to downright visionary. A fine selection of bright-minded geniuses I have encountered on various fora that discuss such questions, or who shared their perspectives on these questions with me as a result of reading my books.

This kaleidoscopic broth of views will depict a variegated map of the human mind-scape, which will allow you to orient yourself and find your own way to the mental destination that fits in best with your convictions. But you are warned as well. As said before you can still take the blue pill and avoid the dazzling tumbling down into the rabbit hole. Your beliefs may be shattered, your faith crushed.

As editor of this anthology, I consider it my duty to set up a framework in which each of the above mentioned perspectives can be hosted in a logical following order and to provide you with a thread and map to find your way on this mind boggling and perilous journey into the abyss of existence: Into the realm of discovery of imagination, apparent emptiness or purported consciousness; yes, of a tower of turtles or a hall of insubstantial mirrors at the foundations of what you used to call "reality".

This is an anthology written by authors from different cultures and backgrounds. As a consequence not each of the articles may appeal to you. In some cultures people want a resume at the start; a succinct distillation of the essence of the book, what it means, why it is important and how we can practically apply its teachings. Other cultures are more theoretic-deductive. They prefer a thorough explanation of the underlying principles before any conclusion can be drawn. Yet other cultures are more contextual and like to have different examples sketched in prosaic, poetic or artistic ways to get a holistic picture of the mindscape, so that they can draw their own conclusions.

I hope this book can satisfy all of these criteria, although it seems like a heroic quest if not a hopeless job to achieve. I therefore apologize beforehand for the inconvenience if you encounter a chapter which does not fit your cultural preference. "If you don't like it, skip it", is my advice. The next chapter might be right what you were looking for.

In this introductory chapter I'll address a number of philosophical and semantic questions to build the aforementioned framework. Thereafter I shall give a very brief description of the contributing authors and a short summary of their vision on this subject-matter. You can then decide for yourself which parts you'd like to explore and unfold.

Theoretical Framework

In order to be able to answer the question "Is Reality a Simulation?" we first have to understand what we mean by the different terminologies used in this question. This is probably a bit dry subject-matter and there is no absolute need to dig your way into the rabbit hole through this hard-core theoretical background. The rabbit-hole network has several entrances, so feel free to jump ahead.

In this theoretical framework we'll have to know what is meant by the terminologies "Is", "Reality" and "Simulation" in the different perspectives of our writers. These are not trivial questions, as we are about to find out.

I start this analysis with the terminology "Reality". Wikipedia gives the following definition of Reality from the Oxford Dictionary, which, as I will explain, is certainly not a definition everybody will be able to agree on:

"**Reality** is the state of things as they actually exist, as opposed to an idealistic or notional idea of them. Reality includes everything that is and has been, whether or not it is observable or comprehensible. A still broader definition includes that which has existed, exists, or will exist."

So Wikipedia considers reality as "things as they actually exist". This brings two further questions, what are "things" and what does "exist" mean. If you click on "exist", you get the definition of existence as follows: "Existence, in its most generic terms, comprises the state of being real", which leads to a kind of circular reference. Then it continues "and the ability to physically interact with the universe or multiverse". The universe is defined as "all of space and time and their contents, including planets, stars, galaxies, and all other forms of matter and energy.

So in other words and less succinctly expressed, Wikipedia and the Oxford dictionary describe reality as what is usually known as "objective reality", namely the total of all matter and energy which are considered independent of the observation by conscious subjects.

It is true that for most people this definition is the common sense definition: Reality is the physical objective reality and not your imagined mental concoctions.

Such a definition is appealable for people who have a so-called "physicalist worldview" in the sense that they consider that matter (or energy) is the fundamental substance in nature and that all mental aspects including consciousness are merely emergent epiphenomena deriving from the interactions of matter/energy. Most scientists, philosophers and engineers, maybe you yourself, in the Western societies adhere to such a worldview.

But this is not a universally shared worldview. In Eastern cultures more "idealistic" notions of reality are considered: In Buddhism everything is present in one universal mind; Hinduism in most of its varieties also considers consciousness as fundamental and matter merely as an epiphenomenon of consciousness! This is your worldview upside-down unless you belong to this daring paradigm threatening congregation of ethereal adepts.

Strange observations in quantum mechanics such as the wave-particle duality observed in the so-called double slit experiment, where the outcome depends on the experimental set-up, have led certain

13

physicists to adopt an idealistic worldview. They consider that the notion of "*if you change the way you look at things, the things you look at change*" implies that consciousness is somehow involved in the collapse of a wave into a particle.

Physicalists on the other hand will wave this argument away, by stating that this is due to interactions with the devices in particular the detector or another instrumental artefact.

In any case there are a growing number of dissident scientists, who question the notion of a fully objective reality. The subjective influence may play an important part in what reality looks like. The idealist Bernardo Kastrup[1] speaks of the elements of reality we agree on not as "objective reality", but rather as a "consensus reality". We assume that people around us have the same types of experiences due to the principle of "Organizational Invariance" of the brain. Because our brains essentially have the same physical form, we assume they work identically and usually indeed we seem to agree on the experiences we have.

So the notion in Wikipedia that "reality is merely the state of things as they exist," is not a notion every idealist will be able to accept. For if reality is an expression of mind, it INCLUDES the notional idea of the state of things. Not your notional idea of things, but that of an all-encompassing kind of mind substrate (which some of you will be tempted to call the mind of God). What the physicalists call "objective reality" might be a solipsistic subjective reality of the all-encompassing mind or it could be a panpsychic interference pattern caused by the interaction of countless small subminds like you and me.

Because among the authors of this book we'll encounter proponents from both sides, it is important to point the differences out to you.

Nevertheless, even for the idealist, there are still parts of "reality" which are at least experienced as an objective reality. It might be that my mind makes an almost infinitesimally small contribution to the wave collapse resulting in the keyboard I am using to write this chapter, from my experiential perspective it is as if this object exists totally independently of me. Whether this is true or not is not a question I am going to answer. I'll leave it up to you to draw your conclusions after having read this book. But we can perhaps agree, that from our human perspective there are parts of reality, which we experience as if they are independent from our observation, which henceforward I will call the "ontic". But there are also parts of our experience, which are experienced as exclusively the domain of our own minds: our thoughts, our imaginations, our dreams etc. and these parts I will hereinafter call "epistemic".

In physicalist philosophy ontic refers to real or factual existence (i.e. objective reality), whereas epistemic relates exclusively to the domain of knowledge. In my definition in this book, the ontic is that part of reality, which we experience as independent from our observation whereas the epistemic is total sum of what we experience as phenomena which are deemed to be more exclusively of the mind.

With these sets of definitions I would like to define total reality as the sum of ontic and epistemic.

Reality = ontic+epistemic   or in other words
Reality = experientially objective reality + subjective experience or
Reality = experientially physical reality + mental reality.

This does not mean that your fantasies are real in the ontic sense, but it means that your fantasies are at least acknowledged as a real thought phenomenon, which is occurring.

With such a quantum-mechanics like superposition definition, we can accommodate both physicalist and idealistic views. What Wikipedia and physicalists call "reality" is merely the ontic, whereas in the eyes of the idealist everything is ultimately epistemic. But the present definition allows us to be able to use one definition for both sides.

The remark of Wikipedia that "reality includes everything that is and has been", is not a notion everyone will agree upon either. In our common sense understanding the vast majority of people is of the opinion that only the present moment contains what is physically there. That's why we say "it is present". Most people do not consider the possibility that everything that was or will be is kept or is made available in some kind of record or multiplicity of parallel realities, where all events of all times that were, are and will be coexist simultaneously, without us being aware of this. Whereas this might be considered to be the case in certain versions of Everett's[2] Multiple Worlds Interpretation (MWI), this is not a version of interpretation all scientists necessarily agree upon. It must be said here that there are certain experiments in physics such as the delayed choice experiment[3], which seem to point to the fact that information can travel backwards in time.

I am not going to state whether this interpretation of MWI is right or wrong, -to be honest: I don't know- but I'd like to make you aware that the definitions you find in encyclopedias or dictionaries do not necessarily represent the truth or the ruling paradigm in science.

From the idealist point of view it is possible to imagine that the past can continue to exist as memories of an all-encompassing mind or consciousness and the future as an exercise of that mind to screen mental possibilities, some of which or all of which it will allow to materialize into physical existence.

From the physicalist perspective the MWI is one logical explanation of the wave collapse in quantum mechanics: Each possibility occurs in a different branch of the universe, which is formed upon the collapse and these parallel realities do not interact.

So reality might be of the present moment only or might include all possible instances of time.

Noteworthy, if we look deeply into matter - which we normally consider "solid" and we experience as "objective reality"- what we discover, is that it is essentially empty. An atom has a core of a few femtometers, which length is 20.000-100.000 times smaller than the atom itself. As the volume of a sphere is proportional to the third power of its radius, we can get a difference between the volume of an atom and that of its nucleus of up to almost 10^{15}. In other words the atom is at first glance already up to 99.9999999999996% empty. The few electrons surrounding the core have a negligible mass and size and can behave as a wave as well, so that their "material nature" becomes a matter of semantics. The core itself is made of neutrons and protons, which in turn are made of subatomic particles like quarks and gluons. Quarks and gluons are considered not to have any or hardly any volume (the radius of a quark has been estimated as smaller than $0.43*10^{-18}$ m. So matter is really, really very empty. What is left is difficult to describe; some kind of localized whizzing energy; a vortex of virtual nothingness; an interference pattern of "energetic waves" or "probability density". Yes, you hear it well; ultimately you are a probability density of a mathematical wave function. Metaphors and descriptions that lose their meaning at these scales, as these properties are not directly observable anymore but can only be inferred from data. To only reason we cannot walk through each other is because of the so-called "Pauli-principle", which states that two particles (like electrons) cannot occupy the same spatial volume[4]. An observed fact, which does not really make us understand why this is so.

Funny enough, it seems we have a kind of cultural prejudice, that only the material world is real. At school my son learned "matter is that which <u>everything</u> is made of" when he was only eight years old. I happen to think there is more to the story. In chemistry when an electron is spread as a wave over multiple atoms, this is not only called delocalization of the electron, but even **derealization**. In other words, society indoctrinates us that only the materially tangible, the localized and particle based forms are real, but energy waves and electricity - although we use them - are considered to be less "real". Yet our whole wireless technology is based on broadcasting non-material electromagnetic energy waves. Somehow in between "materialized states" the electromagnetic energy is capable to retain some structure to convey information. Why should that be considered unreal, metaphysical or ghost-like? Just because it is not localized, particularized? As it influences our material reality, shouldn't it be included in our definition of reality? Is it not that by being part of our technology; by being applicable in a meaningful technical sense, this non-material energy has become part of the physical or of the real? Is it not time to update our terminologies and school books?

To get an idea of the emptiness: Imagine you are a subatomic particle flying through an atom. Then your journey through the atom is like that of a spaceship through a solar system, in which its sun as metaphor for the nucleus would be 10 times smaller than our sun. If you as subatomic particle would be travelling through the human body, this human body would be about the size of the galaxy ($9,46*10^{18}$ m). Maybe you should use the improbability drive[5] from Douglas Adams' Hitchhiker's Guide to the Galaxy" to find your way through this probability density interference pattern.

You wouldn't be able to make much sense of it, just as we can't make much sense of the cosmos. So what is there so real about point-size particles in an empty space? Only if you could discern a greater structure in this grid, you could give it a name or meaning.

If we pick one grid or scheme, reality looks ordered in one way, if we pick another grid or scheme it appears ordered in a different way[6]. Depending on your grid or scheme, you see what you want to see. As R.A.Wilson[7] used to say: "What the Thinker thinks, the Prover will prove". There is a strong cognitive bias when we are trying to figure out what something we encounter actually is. It is only after more detailed inspection that we can discriminate the snake from the rope -if we can rely on our senses. Therefore the Discordians[6] say that Reality is the original Rorschach (The Rorschach inkblot test is a test to see how someone will interpret the shape made by an inkblot and is often given to prisoners to test if they are criminally minded).

In fact today in artificial intelligence we are developing many different kinds of ways to recognize patterns in data. You can use support vector matrices, neural networks (of which there is more than a dozen different architectures around), random forest algorithms, k-nearest neighbor algorithms etc. Depending on which algorithm you choose, you may get quite different results. These pattern recognition algorithms are like the grids I just described. We observe reality in a certain way, because our neuronal architecture allows for a specific kind of pattern recognition, data classification and data clustering. With another grid, you get a different interpretation. So why would our way of interpreting signals be the ultimate reality?

Besides, our senses can only observe an almost infinitesimal small slice of the electromagnetic spectrum (the visible range comprises wavelengths from 380 to 700 nm). With our instruments we can extend that range to gamma rays (picometers) and radiowaves (up to 100 kilometers). But the wavelengths probably continue on both sides to the infinitely small and infinitely big.

We constantly internally hallucinate a "supposed world out there". We strongly filter the overflow of data entering our senses and create a coherent picture therefrom, which may or may not have a certain degree of isomorphism to the ontic reality. This is especially evident when we are asked to focus on a particular activity. If you are asked to count the number of times on people throw a ball to each other, like most people you will totally miss the man in a Gorilla suit walking among the players, because this irrelevant information is filtered out by your brain. In other words, from the sensory data we receive, our brains concoct a simulation which is as meaningful as possible under the given circumstances. In that sense our experienced reality is almost certainly a simulation and consensus reality a kind of collective hallucination.
What is then reality, what is truth? Is that, what is the matter, some kind of Cosmic Imagination?

There is also another definition of reality by Chris Langan[8], who considers that reality includes everything that is real i.e. can influence reality. If it does not influence reality it is per definition not part of reality. If it does influence reality, it must be included per definition. In this definition, there is no outside of reality, there is no "transcendent" dimension, for if it is real it has an influence on reality and is thereby included. In other words, in Langan's definition reality as a whole cannot be simulated - although levels thereof might be. We may consider our "reality" as a simulation, but the world of the simulators is included in Langan's definition of reality too. What is real or virtual then becomes a matter of perspective but this tautological trick does not explain us what reality is.

Perhaps reality is a relative concept. Perhaps there is only one root reality from which a vast array of simulated virtual realities emerged, which in turn created their own virtual realities. Perhaps "real" is only real enough, if reality has predictable laws allowing to create long lasting structures. Perhaps the definition of whether you can consider something "real" depends on whether you can observe it long enough. Perhaps this depends on your "sampling rate". It seems most humans do indeed have similar sampling rates, allowing them to discern the same types of structures in a recognizable manner, which they can relate to each other. Perhaps we should aim for a practical definition of "real", like: "Something is real, if it is real enough to be effectively recognizable by a significant group of people and if it is effectively usable".

Every generation has its own ultimate metaphor for what everything is like. In the seventeenth century everything was considered to be a mechanical clockwork, in the nineteenth century a steam engine, later on electricity played that role and in the twentieth century we saw the rise of the cybernetics metaphor. Yet all of these were defective in a certain sense that they failed to be universally applicable. In the present computer era, everything is considered to be a computational process; an input-output process of binary content. But Turing showed us that computability has its limits. Not only in the per se sense, but also in the practical sense: there are plenty of problems which cannot be solved in polynomial time and the exponential time alternative most often renders the problem practically intractable. But computation is more than binary computation only, although Turing was of the opinion that any computational system can ultimately be reduced to a Turing machine. A Turing machine consists of an apparatus which can read and write digits on a tape and a semi-infinite tape with ones and zeros that can be read and overwritten. Turing did not believe that more complex structures could add anything, which a Turing machine could not ultimately compute. Even many present day neural networks can be Turing machines, but not all of them. So if it turns out to be true that everything physical is the result of a computation-like process, then the term computation needs to be interpreted broader than Turing-computable alone. Perhaps then the ultimate metaphor for reality is the neural network. Maybe everything at every level is either a direct structure of a neural network or it is an auxiliary aspect thereof (like the current; neurotransmitters or other signal molecules etc.). Certain mystics do claim to have seen a tree-of-life like fractal of neural networks as the essential structural and functional conformation of reality[9].

The complete brain architecture is encoded at the level of the genome that functions as a neural network of its own kind. Metabolic pathways and protein networks can be considered as neural networks. Fungi build a network of mycelium allowing trees and plants to communicate with each other. The internal processes of molecules and atoms are eerily reminiscent of neural networks, as are superclusters of galaxies. And then there is feedback between the different levels of neural networks. Certainly epigenetics shows how important influences at the molecular level are cascaded "up" to the neuronal levels in the brain. Perhaps each level of a neural network could be considered as a simulation or a simulator or generator of another neural network level in the physical world. Ultimate reality is then the functioning of a living self-organizing neural network principle that expresses itself from the subatomic to the universe level. Evolution as a learning process from the inorganic to the astronomical.

So with this information we can wonder whether the ontic is a simulation, whether the epistemic is a simulation or whether both are simulations. Or whether in one simulation multiple simulations are nested, if you use a different pattern recognition grid.
We can moreover wonder whether the term Simulation allows for an endless plethora of all possible scenarios.

Most authors in this book however will refer exclusively to the ontic when they speak about "reality", either as "objective reality" or as "what is experienced as objective reality". But in this book, in which we wonder whether ontic reality (as well as epistemic reality) is actually a product of a simulation (which can be a computer or a mind-type simulation), it is extremely important that we are aware of this nuance.

And this brings us to the term "simulation"
Simulation comes from the Latin verb "Simulare", which means to make like, imitate, copy, represent, feign. Its root is the Latin word "similis" which means "similar to". So a simulation is to make

21

something similar to something else. What is this else? It is a supposed ultimate reality. That's why "simulation" is normally defined as the imitation of the operation of a real-world process or system. Today this word most often refers to computer simulations of processes or structures occurring in the ontic world.

As W.H.Tsang[9] describes: "The idea of Virtual Reality is based on mathematical formula and abstract data structures held in the memory of a computer, which allow the computer to render images onto a computer display or virtual reality headset the intricate details of the virtual world being explored by the user, either in some computer game, flight simulator or other VR immersive environment".

Many - if not most- people only think of von Neumann type computer simulations when they hear the term "simulation". Some authors in this anthology indeed use this meaning. But in reality, the term is much more encompassing.

So if we ask whether reality is a simulation, we hypothesize that there might be a more real reality, this reality is similar to. Our reality could be an image of an original object or subject beyond our dimension. It could be an image at a lower resolution of an image with a higher resolution. There could be a whole series of such images all having the function of being both a simulated image and a simulating image.

If we consider a God created or dreamt our experienced world and our experiences in analogy to his own world, we could call that a simulation.

If we consider a computer engineer in a more advanced dimension created our world as a computer model, we could also call that a simulation.

If reality is a self-modifying system, it can be considered to take itself as a model input, process this input and provide a modified version of itself as the result, the model output. This is what we shall refer to as "self-simulations".

So whatever mechanism has been used to create "our reality," it provides an image of another reality or of a prior state of itself.

With the explanation of these various forms of "simulation" it will be easier for you to follow the different authors and you'll be able to see what specific kind of simulation they have in mind when they use the terminology "simulation".

Finally, I'd like to address the terminology "IS".

"Why?" you ask. Do you question our mental abilities? Are we not capable of understanding language?

Well, the term "is" or the verb "to be" is not as unambiguous as you might think.

Firstly, the term to be can be used in a literal sense to define something or in a metaphorical sense to describe something in terms of something else.

Definition types of "to be" are mostly expressions which show that the term to be defined is a specific instance of the general class mentioned in its definition. E.g. "a car is a vehicle", "a parrot is a bird", "he is the president".

Metaphorical types of "to be" are expressing a likeness, a similarity. For instance, "your son is a giant" or "this meeting is a disaster". What you mean is that the meeting is like a disaster; your son is like a giant. What I consider remarkable in the culture of American youngsters, is that many of them use the terminology "like" at any opportunity, regardless whether a literal sense or a metaphorical sense is intended. Is this a sign of linguistic decline or are we witnessing semantic drift?

Both these types are also known as the "predicative nominal over the subject".

To be can also express a characteristic of the subject: "these roses are red"; "the food is delicious". In grammar this is called a "predicative adjective over the subject".

Furthermore, "to be" often expresses a permanent state. The question is, whether this sense of "to be" can apply to reality at all. As far as we know on the basis of our observations, there is no single thing in reality that remains unaltered. Everything is in a constant flux. This is why the Greek philosopher Heraclitus already said that you cannot put your foot in the same river twice. Even atoms are ultimately destroyed in a black hole or big crunch of the universe if one day it starts to collapse, or are ripped apart if the universe continues expanding and ends in a big rip.
Our bodies are in a continuous state of turnover. Every seven years or so every atom in your body has been exchanged. So what does "is" mean, if the same object is no longer there the next moment?

The verb "to be" is an oddity and invention of the Indo-European language family[10]. Other language families do not even know this word, because their languages are in a so-called "rheomode" or a streaming mode. They rightfully acknowledge that the ontic reality knows no eternally stable structures.

Ontology building

In my book "Is Intelligence an Algorithm?"[11] I indicated that a pragmatic approach to creating a full-fledged ontology of a concept was to address the 6W (What, Who, Where, When, Why and hoW) questions, which we can now formulate for the concepts of "Is Reality a Simulation?"

So what do we mean when we say "Is Reality a Simulation?" Do we mean that reality is like a simulation or do we mean it literally? Do we mean that reality was started as a simulation and then allowed to develop itself further or is it an ongoing simulation process? Is it a dream-like simulation, a computer simulation, both or something else altogether?

What is simulated and why? Is this an ancestor simulation of a post-singularity society that tries to reverse engineer its history to figure out how their singularity came about? Is this a mere screening process of crazy God-like entities, who wish to figure out what are the conditions for the emergence of hyper intelligence, which is capable of recruiting all available energy sources for its own benefit? Or is it a less spectacular Petri-dish like experiment we're in? Is this theatre a mere entertainment spectacle performed to amuse our overlords? Is it a physics and/or biology experiment? Are our overlords trying to find a reason for their existence themselves via this simulation? Are they playing a game and are we mere puppets or marionettes? Are some of us incarnated overlords playing the game and others non-player characters (NPC) devoid of consciousness? Did we volunteer to enter this game by consenting to have our entire pre-game memory erased? Or are we spirit souls on an endless journey of life and death damned by the curse of physical reincarnation? Are we the mere thoughts of a God-like cosmic mind?

And who do we consider to be the designer of the simulation? Is it a computer in another dimension? Is it God? Is it a Kardashev IV society? (The Kardashev scale measures the technological advancement of a civilization; a scale III is capable of controlling all the energy of a host galaxy. A scale IV has no official definition, but I baptize it as being able to control the entire energy of a universe including the ability to create simulated universes). Is it humanity itself in the far (or rather near) future of the Singularity - or what Frank Tipler[12] considers the Eschaton or the Omega hypercomputer at the end of time?

Where is this simulation performed? Is it on the surface of a Black hole computer that generates our world as a 3D hologram? Is the Universe one big neuronal network and are we its mere thoughts? In the Mind of God? In the cat litter of the Simpsons? Is there an exit out of this rabbit hole of material entanglement? Is this Hell; is it Heaven, the Cyberbardo or all of the above? Is it a set of nested simulations? How many are there? Is there a ground zero reality? Or is it turtles all the way down?

Is it performed at the end of time in the Omega point? Or even beyond time? Or is it simply a time creating process? Are things rendered only when we observe and/or measure them? Is this the rationale behind the Heisenberg uncertainty principle? Is it performed as a linear sequence of series of simulations all starting with a big bang and ending in a big crunch or big rip? Or are the simulations performed in parallel to give rise to Everett's Multi World Interpretation?

I leave it up to the authors to use the definitions and approaches they consider appropriate and I am sure that after you have read about the possibilities of all these nuances, you will be able to grasp the intended meaning from the context.

This book will be structured as follows: It will start with the contributions of those authors who consider that reality is indeed the product of a simulation. It will be followed by the antithesis of those authors who consider it's not. And it will end with a kind of synthesis of a reality which is neither simulated from a different dimension nor the product of merely physicalist mechanics, but rather the product of a self-modifying self-simulating principle.

Chapter 2

The Historical roots of the Simulation hypothesis

By Antonin Tuynman

It is hard, if not impossible to determine how deep the roots of belief in the tree of virtuality reach back in the history of the human race. Certainly we did not need our Oculus Rift VR headset to come up with this idea. When did man first started to doubt the reality of the solid world around him? When did we first start to think we might be living in an illusion?

The first written tradition is perhaps the allegory of the cave by Plato[13] in his treaty "The Republic" written around 380 B.C.

In this allegory, prisoners are chained in a cave in such a way that they can only look at a wall in front of them. Behind them is a fire burning and between the fire and the prisoners is a low wall, behind which other people walk carrying objects or puppets "of men and other living things". These objects cast shadows on the wall in front of the prisoners. The sounds made by the walking people also echo from this wall, so that it seems that the shadows are making these noises. For the prisoners, who never experienced anything else these shadows are the only reality there is. In the story one man escapes. At first according to Plato he would not understand that what the prisoners see and hear are mere shadows and echoes. Only once the escaped prisoner found out about sunlight outside the cave and became accustomed to it, would he be able to learn about shadows. Thus he'd start to understand the real source of the images and sounds. Thereafter he'd consider this new reality superior and "he would bless himself for the change, and pity [the other prisoners]" and would want to bring his fellow cave dwellers out of the cave and into the sunlight". Unfortunately back in the cave his eyes would need to get accustomed to the dark again. His fellow prisoners would think he'd gone blind and conclude that it's dangerous outside of the cave. They would not be willing to leave.

As the freed prisoner in this allegory represents the person who sees the world for the illusion it is, this is one of the first written proofs that humans were conscious of the possibility that our reality might not be the ultimate reality.

Plato was also the father of ideas of like a Cosmic Mind or World Soul (psyche kosmou) and a transcendent cause or Intellect (Nous). The makes him one of the first idealists. What really existed for Plato were the "forms" or "ideas" which are the essences or real natures, of which the concrete things of appearance are merely the imperfect reflections. The work of a skilled craftsman he calls the Demiurge. This idealistic worldview would remain the default paradigm until at least the mid nineteenth century.

The earlier philosopher Pythagoras (circa 500 BCE) already stated that the universe was numbers. We can wonder whether he had a mystical anticipation of digital physics.

The Greek sophist, Gorgias (c. 483–375 BC) is reputed as the father of Solipsism, the notion that we can only be sure of our mind to exist. Moreover he's quoted[14] to have stated:

1. Nothing exists.
2. Even if something exists, nothing can be known about it.
3. Even if something could be known about it, knowledge about it can't be communicated to others.

With this reasoning the Sophists tried to show that "objective" knowledge was a literal impossibility. An extreme interpretation of Solipsism is to assume that only I exist and that everything else is a concoction of my mind. Or a simulation, if you wish.

Another early text on this topic is from the Zhuangzi[15] by the eponymous author who lived between 369 and 286 B.C:

"Once upon a time, I, Chuang Chou, dreamt I was a butterfly, fluttering hither and thither, to all intents and purposes a butterfly. I was

conscious only of my happiness as a butterfly, unaware that I was Chou. Soon I awaked, and there I was, veritably myself again. Now I do not know whether I was then a man dreaming I was a butterfly, or whether I am now a butterfly, dreaming I am a man. Between a man and a butterfly there is necessarily a distinction. The transition is called the transformation of material things. "

Dreaming in fact is our most direct springboard to question whether our reality is an illusion.

Both Vedic and Buddhist traditions have spoken of the world as Maya, a magic or illusory veil. Maya has been said to be the reflection of something very real in a spiritual world. The powerful and colorful paintings the Tantric and Tibetan Buddhists have used were intended to help them visualize alternate realities. The so-called Avatamsaka Sutra[16] from about 100 B.C. speaks of "infinite realities". In images we find an enlightened deity sitting calmly on a lotus flower often in the middle of raging fires. Worlds within creatures and worlds within circles, representing the karmic cycle, show how we are caught in the web of dependent arising. Fortunately, there seems to be a way out of this Maya. A little rainbow colored path leads the enlightened ones to the realms of the deities.

Even today, certain schools of thought in Buddhism teach perceived reality literally as unreal. Chögyal Namkai Norbu[17] considers all our sensory perceptions as a big dream.
From a neuroscience perspective he is actually right in a certain way: As explained in the previous chapter, when you think you see the outside world, actually what you are experiencing is an image concocted by your brain.

Interestingly, in the Tibetan Book of the Dead[18] an intermediate state of existence between two lives is explained, which is called the Bardo. This is a dream-like state in which various realms are visited before a new incarnation can occur. Some psychotic patients believe they are already dead (Cotard Delusion) and believe they are traveling through this Bardo. A computer game in greater network game Second Life is also based on travelling through the so-called "Cyberbardo". Perhaps the dead-alive dichotomy is a false one and only journeys through a Cyberbardo simulation exist.

In Hinduism[19] we also find the notion that Vishnu, the all-pervading one, lies in a dream state on the serpent Adisehsa Ananta. Ananta is time and floats for eternity on the ocean of Cosmic Consciousness. Brahma is born out of the navel of Vishnu and begins the process of creation. Vishnu expands into everything thereby becoming everything. By the act of watching his dream, including the creation of the universe by Brahma, Vishnu sustains the Universe. Only when Vishnu wakes from his dream, does the cycle of creation end.

In the Hellenic world we find the notion of Hermes Trismegistos, the thrice great one. This Godhead seems to be a merger of the Greek God Hermes and the Egyptian God Thoth. The earliest texts mentioning this God go back as far as 172 B.C. Hermes Trismegistos is reputed to have written the so-called Tabula Smaragdina[20] (which may have an origin much later; it's earliest written version is an 8th century Arabic text). This Tabula Smaragdina mentions the concept "As above, so below", which seems to be considered as an absolute truth among esoterically oriented people today.

"That which is Below corresponds to that which is Above, and that which is Above corresponds to that which is Below, to accomplish the miracle of the One Thing."

The concept "As above, So below" implies that our physical world is a reflection of a spiritual world. That the microcosm (oneself) is similar in structure to the macrocosm (the universe).

Around 200 A.D. we find the sect of the Gnostics. Gnosticism[21] is a peculiar religion, which describes a dualistic cosmos. Spiritual sparks or souls have become trapped in matter but can be freed by saving knowledge or "Gnosis". The world is the creation of the Demiurge Yaldabaoth, which literally means "Child, come hither". However in another translation his name is translated as "Child of Chaos". The gnostic myth recounts that Sophia (literally "wisdom", the Demiurge's mother) desired to create something apart from the Father to which he did not consent. In this act of separation, she gave birth to the Demiurge. Being ashamed of her deed, she wrapped him in a cloud and created a throne for him within it. (Note that with this "Demiurge" not necessarily Plato's Demiurge was intended).

As the Demiurge of Gnosticism did not see Sophia or the Father, nor anyone else, he thus concluded that only he himself existed. In a quote from the Apocryphon of John he is reputed to have said: 'I am God and there is no other God beside me'.

However, he did not know the source of his power and did not know that there was someone above him. Keep this in mind when you read the chapter on "Pandeism and Simulation theory" by Knujon Mapson[22].

The myth furthermore shows how this first separation later on resulted in the entrapment of the divine spark, Sophia, within the human form. This spark is latent until awakened by a call, and knowing oneself as this divine spark is the beginning of restoration of Sophia, as well as gnosis.

It gets a bit blurry when we see that in Gnosticism, Yaldabaoth created the world together with six other so-called Archons. One of these is "Sophia" (Venus), whereas Yaldabaoth himself is indicated to be Saturn.

The internet is full of conspiracy theories that somehow link the Gnostic teachings to the notion that we live in a simulated virtual reality ruled by these evil Archons, who feed on our energies. Many of them claim that one of the Archons is called "HAL", which a Coptic word meaning "simulation" and that this is linked to the "HAL 9000" computer in the book and eponymous film "Space Odyssey 2001". Neither Clarke[23] nor Kubrick has confirmed this assertion. Rather Clarke indicated that this name derived from "Heuristic Algorithm". Another popular theory, though denied by Clarke and Kubrick, is that Hal is actually a sly reference to IBM (made by moving back one letter from each letter of that acronym). I have not been able to find this name Hal in the Apocryphon of John[24] either or in any other texts of the Nag Hammadi English Library[25].

These theories rather feed on themselves, braiding an unintelligible and inextricable knot of nonsense. In my humble opinion it appears to be a 20th century myth invented by a self-educated and self-proclaimed scholar named John Lash who twisted facts to feed a hungry audience of conspiracy theory addicts.

In the middle ages we encounter a number of theories that reality is made for us as a kind of "cosmic intelligence test". In the book the "Assassins of Alamut" Anthony Campbell[26] refers to the medieval Islamic Sect of the Ismailis". They believed the Koran contained an esoteric secret, to which they held the key. He claims that similar notions can be found in the Kabbalah, the teachings of the Cathars of the Languedoc and in the Gnostic texts, all relating to the theme of the world as an unreal simulation, veiling an unknown ultimate reality.

Kabbalists reveal in the book of Zohar[27], that what is revealed to us as matter is merely one thousandth of the total matter and that there is nothing real about it.

In the early renaissance we can see that the idealistic paradigm was still firmly rooted in, for instance, the painting "The Garden of Earthly Delights" by the Dutch painter Hieronymus Bosch. This "tryptich" (three-panel) oil painting shows on the left hand the paradise or "garden of Eden", where ideas are present in a pure and singular form. In the middle panel, representing "Earth" we see a lustful humanity and other manifestations in multiple instances. In the third panel, "Hell", forms are not only multiple, their logical cohesion is compromised by absurd combinations of parts of the pure essences or ideas. We find similar notions in the Divine Comedy by Dante written in the late middle ages.
In the 16th and 17th century Galileo[28] would write that the book of nature is written in mathematical language and Newton would call God "a mathematician". The notion that the physical universe is ultimately a mathematical construct implies it is a kind of simulation.

In the 17th century René Descartes[29] came with his famous "dream argument". Dreaming according to Descartes is proof that we cannot fully rely on our senses to distinguish reality from illusion: "Whatever I have accepted until now as most true has come to me through my senses. But occasionally I have found that they have deceived me, and it is unwise to trust completely those who have deceived us even once."

His contemporaries Locke[30] and Hobbes[31] have tried to refute his argument by stating that in dreams there is no pain and that dreams are susceptible to absurdity unlike waking life. From my personal experience I disagree with this counterargument. As a child I have worn braces on my teeth and sometimes in my dreams I still feel the pain thereof, whereas in waking life I have been free of these braces for over 30 years now. In addition, not every dream I have is necessarily absurd. Some dreams are eerily realistic.

Descartes even went so far to suggest that even his body (as well as everything else) perhaps only existed as an idea in his mind, thus venturing on the path traced by solipsism. Eventually, he concludes that there is a mind-body dualism, but at least he explored the possibility of "ontic reality" as a mind simulation.

Kant[32] was also of the opinion that we could only know the appearance of things (phenomena), but not the things in themselves; the world as it actually is, which he called the "Noumenon". In other words our mental image of the world is a mere simulation.

In the 19th century the theosophy movement of H.P Blavatsky[33] introduced the notion of the so-called Akasha, borrowed from the Vedas, which she referred to as indestructible tablets of astral light. This notion was picked up by Alfred Percy Sinnett[34], who describes a Buddhist belief in a permanency of records of everything that has happened in the Akasha, as well as the ability of man to read the same. This notion has developed to the so-called "Akashic Records", very popular among the esoteric people. Interestingly, this concept of the Akasha has in the 20th century been equated with the so-called quantum vacuum, quantum foam or zero point-field, which according to certain adepts can function as a memory and a digital computer. The world we live in is then nothing else than ones and zeros on an Akashic switchboard, which can either be a simulation carried out by entities from a higher dimension or a kind of "self-simulation".

Most of these theories have a religious connotation. And this is not surprising. After all, every notion that a God or Gods created or designed the world can be considered as a simulation argument. God as the grand architect of the universe.

Noteworthy, the German idealism of Hegel, Fichte, Schelling and others[35] can be considered as one of the last major convulsions of idealism philosophy. The "Ding-an-sich" or "thing-in-itself" is here a more modern version of the Aristotelian essences.

In the 20th century Pierre Teilhard de Chardin[36] published his extraordinary visionary book "The Phenomenon of Man." Written in 1938, it predicted the advent of a so-called "Noosphere", a layer of knowledge covering and connecting our planet, which has found a physical expression in the form of today's internet. De Chardin described that evolution has a direction, namely the direction of concentrating consciousness in form, striving towards an apotheosis of knowledge, which gradually is attained by the formation of the Noosphere and which will culminate in the theogenesis of the Omega point.

Later Frank Tipler[12] equated this Omega point with the Eschaton; a universal computer generating our (and other) realities as a simulation. Eschaton comes from the Greek word eskhatos, which means "the end". The Eschaton is the last thing, the final thing. In Tipler's and McKenna's[37] opinion, it is also the ultimate only thing, namely a universal computer generating countless virtual realities with instances of itself. In other words, we're all an instance of God. You, I and everybody else is God, exploring itself.

The late 20th century and beginning of the 21st century is the cradle of digital physics: The theory that the whole universe is describable as information, which can be conceived as the output of a computational device. It is not very remarkable that the idea of simulation theory is popular among its proponents. In his book "Rechender Raum" (1969) Konrad Zuse[38], the inventor of the programmable modern computer, was the first to hypothesize about the universe as a digital computer. Edward Fredkin[39] coined the terminology "digital physics" and later "digital philosophy" (2003). His ideas join those of Stephen Wolfram[40] who sees the universe as the output of "cellular automata", a modern version of Leibniz' Monadology[41]. In the same breath we can refer to John Archibald Wheeler's[42] "It from bit", Max Tegmark's ultimate ensemble[43] and to Carl Friedrich von Weizsäcker's binary theory of ur-alternatives[44], pancomputationalism and computational universe theory. Modern quantum based variants have been proposed by Paola Zizzi and Seth Lloyd. The Dutch physicists 't Hooft[45] (holographic principle) and Verlinde[46] (entropic gravity) both concur with the notion that the physical universe is made of information, of which energy and matter are merely manifestations. In the holographic principle, which is a principle of string theories and a property of quantum gravity, the description of a volume of space is said to be <u>encoded</u> on a so-called lower-dimensional boundary to the region—preferably a light-like boundary like a gravitational horizon of, e.g., a black hole. This urges us to ask questions like "Are black holes computers?" and "Is the underlying source of our apparent "reality" a network of interlinked back hole computers?" Noteworthy, deep in the equations of supersymmetry the physicist James Gates[47] found what is essentially an error correcting computer code. In contrast, Bruno Marchal[48] argues against digital physics. He starts from a different premise and concludes the presence of a so-called "Digital Mechanism", in which arithmetic constitutes the ontological primitive, from which the laws of physics can be derived.

In the "Fabric of Reality" David Deutsch[49] ties together four strands of knowledge -Everett's multi world interpretation, Popper's epistemology, Turing's theory of computation and Dawkins' evolutionary synthesis- in an orgiastic broth, resulting in the notion of "reality" not only being the result of a multiversal quantum computing process, but being the multiversal quantum computer itself. Not in the least this notion is founded on the Church-Turing principle which states that "every finitely realizable physical system can be perfectly simulated by a universal model computing machine operating by finite means".

The 20th century is full of fictional literature, lecture and movies about dream and otherwise simulated realities. Most of these are of the cyberdystopian genre.

It would go beyond the scope of this chapter to give a comprehensive overview of these. I just picked a few author names for illustrative purposes.

An absolute master of science-fiction involving simulated realities is Philip K.Dick. There's even an annual Philp K.Dick prize for the best SF book. Robert Heinlein, Stanislaw Lem, Douglas Adams and William Gibson are other great authors if you wish to explore the VR-scape in literature. Most of these books were written in the second half of the 20th century.

The book "Neuromancer" (1984) by William Gibson[50] deserves a special place in this row, as he's the first to use the terminology "the Matrix" to indicate the global computer network in cyberspace which is a virtual reality dataspace. Unlike in the movie "The Matrix", in Neuromancer people still also live in the "real" world and can access this cyberspace in a full immersive mode. The main character is unable to access this matrix, because his nervous system has been damaged by a mycotoxin, which had been administered to him as a punishment for a crime.

In the movie culture we can go back as far as 1973 where in Fassbinder's "Welt am Draht" (World on a wire) adds the theme of recursivity by having the very people who work on simulating a world find out that they actually live in a simulated world themselves.

The Matrix (1999) is probably the best known movie on theme of a computer simulated world. In the unlikely event you haven't seen it; I'm not going to give away the gist of the film.

This brings us to the 21st century where we have the formalization of the simulation argument by the Swedish philosopher Nick Bostrom[51].

Bostrom's trilemma argument states that at least one of the three following propositions is almost certainly true:

(1) The fraction of human-level civilizations that reach a posthuman stage is very close to zero;

(2) The fraction of posthuman civilizations that are interested in running ancestor-simulations is very close to zero;
(3) The fraction of all people with our kind of experiences that are living in a simulation is very close to one.

If (1) is true, then we will almost certainly go extinct before reaching posthumanity.

If (2) is true, then there must be a strong convergence among the courses of advanced civilizations so that virtually none contains any relatively wealthy individuals who desire to run ancestor-simulations and are free to do so.

If (3) is true, then we almost certainly live in a simulation. In the dark forest of our current ignorance, it seems sensible to apportion one's credence roughly evenly between (1), (2), and (3).
Unless we are now living in a simulation, our descendants will almost certainly never run an ancestor-simulation.

His trilemma reasoning departs from the concept that a technologically mature "posthuman" civilization would have enormous computing power. Even if only a tiny percentage of them were to run so-called "ancestor simulations" (i.e. "high-fidelity" simulations of prior ancestral life that would be indistinguishable from reality to the simulated ancestor), the total number of simulated ancestors, or "Sims", in the universe (or multiverse, if it exists) would greatly exceed the total number of actual ancestors.

Bostrom then uses a type of anthropic principle reasoning to claim that, *if* the third proposition is the one of those three that is true, and almost all people with our kind of experiences live in simulations, *then* we are almost certainly living in a simulation.

Most books about Simulation Hypotheses are endless repetitions of Bostrom's argument. This book is not one of them. It will argue in favor or against a simulation hypothesis starting from other axioms or observations.

Starting with Vernor Vinge's[52] ideas about the upcoming Technological Singularity (a point in human history beyond which our future predictions and speculations become pointless as this Technology explosion will transcend our way of living completely and dramatically beyond compare. Some suggest it may grant us immortality and other Godlike properties if we succeed in uploading ourselves to the singular Webmind, in which we can shape a simulated virtual reality (or a plurality thereof) as our "new reality"), the 21st century is the arena of the Technopapes such as Ray Kurzweil, Peter Diamandis and Elon Musk.

Their visions consider that if we are not already living in a simulation, before 2045 we will be able to upload our minds to a computerized substrate which will run an infinity of computer simulations, which will appear as real as what we now consider to be ontic reality.

A recent study by Ringel and Kovrizhin[53] claims that we are not living in a computer simulation based on their attempts to test an anomaly known as the quantum Hall effect using a technique called quantum Monte Carlo - a computational method that uses random sampling to study complex quantum systems.

The complexity and computing power increase exponentially as the number of particles required for a full-fledged simulation grows. With their technique, to store information about a couple hundred electrons, one needs a computer memory that requires more atoms than what's available in the universe.

To conclude from this that we are not living in a simulation seems rather premature to me. Firstly, the simulation only needs to render what we are looking at and does not need to render the complete universe. Secondly, the computational technique is one chosen among many and may very well not be the best method to study such a system at all. Thirdly, a von Neumann type computer may not be what the universal sim is run on and lastly there may be computational abilities to be discovered below the atom scale.

Chapter 3

Jailbreak – Six Scenarios

Dirk Bruere attended Nottingham University and later what is now Westminster University, and has a BSc in Physics. Subsequently pursued a career in electronics and computer research and is currently employed as an R&D Scientist/Engineer at Surface Measurement Systems.

A founder member of the UK Transhumanist Party, as well as the Social Futurist organization Zero State, emphasizing activism in the areas of society, economics, politics, Transhumanism, religion, and art. A member of the Futurists Board of the Lifeboat Foundation.

Other interests include the interface between technology, theology and the occult explored in the books TechnoMage and The Praxis. Co-presenter of a UK radio show, OneTribe.

For several years held the position of Branch Master in the World Shorinji Kempo Organization, teaching Zen and martial arts, although is now retired from a teaching role.

OK – so we are trapped in a simulation – how do we break out? But before we do, perhaps we should give some thought as to whether this is a prison... or a sanctuary. No? Alright then let's get digging...

First, we have to decide what type of simulation we are in, because a lot depends on the thickness of those walls and bars. So, what are the most likely types of simulation (or Sim) in order of probability, and their scale?

There are various possible degrees of simulation of reality that, although they might appear to us to look the same from the inside, would in fact be utterly different especially in terms of the computing power required to execute them. In addition each type would likely have completely different motivations behind their creation and certainly require vastly different resources. However, before we do that let's eliminate the "cheapest" version, which is the one depicted in the movie *The Matrix*. This is where we are essentially a "brain in a jar" with our senses hijacked and are fed false information. What makes this exceedingly unlikely is the effects of drugs like LSD, which do not merely distort sensory input but which create unique mental states. That could not happen under these circumstances.

Other scenarios we can generally rule out are computer games, at least of the trivial entertainment variety. We appear to live in a lawful world. There are no overt Gods or Supermen flying around subverting the laws of physics at will. Finally, there is no way of determining when this simulation started. It could have been billions of years ago or last Thursday.

So let's start at the one that, for some reason, people assume is "the" simulation.

Scenario 1 – The Universal Simulation

This is the most "expensive", or computationally intensive scenario. This is a simulation at the Planck scale of our universe. For those unfamiliar with the term, suffice it to say that it appears to be the smallest scale possible, being some 20 orders of magnitude smaller than the nucleus of a hydrogen atom (that is, one followed by 20 zeros). It is where concepts such as space and time cease to have any meaning. If our universe is a simulation, and is being simulated at that level of detail there are two conclusions we can draw from this. The first is that the simulation will be absolutely perfect and not detectable. The second is that it must be simulated from a universe where the laws of physics are substantially different from this one, because it is unlikely in the extreme that our universe possesses the necessary computing resources. Given this, there's not much point on speculating further, although of course scientists have done so, most notably Frank Tipler[54] He postulates that in the final moments of a collapsing universe enough energy becomes available to run an infinite number of simulations at such resolution. Apart from obvious problems such as the fact we do not have a suitable physical theory that would allow us to definitively state that it is at all possible, it has subsequently been discovered that our universe appears to be open. That is, it will not undergo a final "*Big Crunch*" as required by Tipler. However, as long as "some universe somewhere" supports such physics then it might be conjectured that our universe is one of those infinite simulations.

Chances of escape? Effectively zero unless we can marshal resources on a trans-galactic scale.

Scenario 2 – The Elder Gods

This is where things start to get interesting...

Consider the notion that the first civilization to arise in our galaxy (the Elder Gods) goes the Transhumanist route and expands to effectively strip-mine its own galaxy which it converts to Computronium (the apocryphal universal computational substrate) in order to maximize resources. As to why they would do this... well, we have barely got into computers and we are already using a significant fraction of our total global energy to power them, and it continues to increase exponentially. Computer power is *the* essential resource for an advanced civilization. Note that this could have happened billions of years ago.

Clearly such an act would curtail the evolution of life throughout the galaxy and certainly block naturally arising intelligence. From the Elder Gods point of view this in itself might be considered a great material loss as well as being unethical. So, how can they both have their cake and eat it? The answer is to run every planet they convert into Computronium in a simulation, most probably "on site" in a small corner of the Computronium that was made from the planet. Evolution would then proceed normally, or at least as normally as could be expected. In the Sim life would arise, then possibly intelligence. And if it followed the same route it would expand through its simulated universe converting all resources and recapitulating the steps of the Gods. However, there would be a number of differences.

The first, and most obvious, would be that only the target planet and its local environs would be rendered in any detail. In other words, from the inside it would look like there is only one civilization per universe. The inhabitants would ponder the Fermi Paradox...

The second is that as soon as the newly arrived locals (us) started converting their simulated locale into simulated Computronium they would suddenly hit a metaphorical brick wall. Because the matter in our universe is only coarsely simulated, the Sim itself would be limited by

41

the amount of Computronium allocated by the Progenitors in the real world, which would be vastly less than the apparent bulk of the planet, let alone the universe. Indeed, it might only amount to a few tonnes. The result would be that as we create our own Artificial Intelligences or became PostHuman the whole facade would break down.

At which point one of two things would happen. Either the Elder Gods would terminate us, or they would invite us to enter the real world. In other words, there would be an unveiling, or revelation, of the true nature of reality. In a very literal etymological sense it would be the apocalypse.

Nevertheless, before this stage is reached there would be a brief intermediate state where a PostHuman civilization of modest means would be feasible. At our present rate of progress it would likely last somewhere between a century and a millennium. During this period we could expect to possess sufficient processing power to run Sims of our own. In fact, there is no reason why Sims could not be nested, with each requiring less processing power than the one above it. The only thing we can state with certainty right now is that if we are living in a nested Sim then ours is right at the bottom.

As for escape, we are building the tools to do just that even as I write. The result being, we will be paroled or executed, depending on our behavior and the likelihood of us being good neighbors when invited to meet the landlord for tea and biscuits.

Scenario 3 – The Ancestor Sim

This is a very popular explanation. The idea is that some future version of Humanity runs Sims of the past for various reasons ranging from historical research of counter-factual outcomes of events (what-ifs) to sociology, economics, demographics and so forth. Equally probably, or perhaps even more likely, is that we are in a Sim run by extraterrestrials who have picked up the TV and radio broadcasts from Earth and are using this as a way to investigate who we are. As such it forms a subset of the Elder Gods view and shares some of its features.

The question of escape becomes one of an escape into an alien reality far from the Earth we know both in time and space. Nevertheless, there are a couple of obvious methods we might try. The first is to render the

Sim redundant by publicizing the fact that we are in a simulation and convincing the vast majority of people that this is so. It's a giant message shouting: "The game is up – we know what you did!", rendering their experiment futile from that point on. As above, it is a serious risk which depends on what they think of us.

It may also be important to try and find, or at least guess, where we really are. Where we are being "run". Given the serious computer power required suggests something extraordinary if it is not cosmic engineering, of which we see no evidence in our sky. So, speculating wildly let me make a suggestion – we are currently located some 27000 light years away in or very close to Sagittarius A* aka the black hole at the center of our galaxy, and the date is somewhat later than 29000CE. Or possibly in one of its tens of thousands of orbiting black holes. Why there? Well, it may be possible to use a peculiar feature of a rotating hole called Malament-Hogarth spacetime[55] in which hypercomputation is possible, which allows near infinite computation to occur in a finite time.

The more subtle way of attracting the attention of the aliens is to send a message directly to the site of where we are, using lasers, radio waves etc. with some kind of embedded message. But, I hear you ask, won't that message take 27000 years to get to the galactic center? The answer is "no" - because we are already there. It is something that I, as part of Zero State, am actually doing as a subset of the METI program (messaging extraterrestrial intelligence). I shall keep you informed of any resulting godlike manifestations...

Scenario 4 – Judgment Day

This is a version of the ancestor Sim. For example, one of the most plausible methods for reconstructing the dead of past ages, or at least us, is from records such as DNA, medical records, photos, writings, videos and so forth. The argument runs that if a simulation of (say) myself could be created such that in the simulation I am writing exactly these words at exactly this time it would be a fairly accurate reconstruction of the historical "me", long dead. However, would that really be true? The counter argument is that it is only a copy, albeit possessing its own life and sentience and not the "real" me at all. I remain dead. To actually be me the copy has to have identical brain states with the original. Unfortunately some fairly crude back of envelope math suggests that the necessary information output from a person is insufficient by at least three orders of magnitude to select one unique state from a possible ten to the ten to the sixteenth power states.

Hence most of the data used in the reconstruction e.g. "what I had for breakfast on 1 January 1990" has to be a guess that, although possibly having an effect on defining my mental state, may not affect the words that I am typing now. Does it matter? When is a copy good enough to truly be "me". The answers are unknown, but my own view is that if I am to be brought back from the dead there should be a subjective continuity of consciousness which an imperfect copy cannot possess. Otherwise, it is someone else—a nearly identical twin, but not the real me.

However, if we live in a sufficient large multiverse (one or more of several different types) this may not matter because somewhere we can achieve perfection in the reconstruction. This is where the multiverse can "rescue" the situation because part of the reconstruction process can randomly guess what should fill the information gaps. The result of making that random guess is a spread of possible versions of the deceased across the multiverse, including at least one that exactly matches the original. So, the entity doing the reconstruction gets a resurrected person that exactly matches all the data they have of the original—which is as accurate as they can get. On the other hand, somewhere in the multiverse a true and perfect copy is produced that has the requisite continuity of consciousness.

Typically, the pertinent questions concern who is running such simulations, how many of them there are, and why they exist. The answers I suggest are a PostHuman "us", billions, and they exist in order to resurrect our dead families. They are necessary in the resurrection process because of a consequence of the Halting Problem, namely that it is in general impossible to jump to the output of a program, in this case a reconstructed personality, without executing the intermediate stages—the life of that person.

This in turn implies that if the above is true it is overwhelmingly likely that we are in just such a simulation. The period we live in is probably unique in being the most information rich in history to date, the one where people still die and also the one that immediately precedes the Singularity. That being so, there are people alive today who will still be alive in the PostHuman transition. However, it is likely that those people will have parents and grandparents who did not make it. Reviving family would, for me at least, be a very high priority. So too presumably for billions of others.

This naturally leads to what may be termed Judgment Day, so-called because some hard choices would have to be made. For example, do we really want Jack the Ripper and Ted Bundy revived and let loose? I think not. Those people would appear to be obvious cases for non-revival, but where is the line to be drawn? There are other options. The final personality could be substantially modified in order to "wash their sins away", or at least their desire to sin. Or if some turning point in their past could be found then they could be set on the road not taken, and "reincarnated".

Which returns us to a question quickly dismissed at the beginning of the chapter – do we really want to escape from what would be our ultimate life support machine? I would suggest not, but...

How might one end this simulation and return to the real world? It has been speculated that there might be hidden codes one could utter, like spells, in order to shut down the program or remove oneself from it. Clearly, speaking the words "End the simulation!"[i] does not work and it cannot be something so obvious. Or can it? Perhaps every one of us already knows the exit code – we call it *death*. However, before you hang, shoot or poison yourself in order to exit a particularly nasty bit of what you believe to be simulated reality, ask yourself this: "Why am I here in the first place, given that I would have inevitably known such a state was not only possible but likely?" Pushing the exit button may well simply result in a "Fail" and you might be forced to resit the exam. Again. Alternatively, allow it to run its course to one of two possible endings. The first is, as mentioned, death. In which case it would be just like waking up in the morning from a particularly vivid dream. The second is more radical, which is to get out alive by riding this Sim's

---

i   Yes, I have actually tried it

45

technological wave to the Singularity where our consciousness expands and merges with the already PostHuman self.

Scenario 5 – The Solipsistic World

Solipsism – the notion that only "I" exist and that everything around me is a figment of my imagination. Computationally, a solipsistic Sim is the cheapest. In other words, only one thing is being simulated in any detail, and that is me. Indeed, if done at the neural level we almost have the power to do it now. If this is so, where does that leave other people? It rather implies that the vast bulk of Humanity consists of what the games world calls NPCs, or "non player characters". They do not have to be simulated in any detail whatsoever except as they interact with me. At the lowest level they would be only be assigned enough resources to pass the basic Turing Test. For example, it does not require much for them to say things like: "Do you want fries with that?" when ordering food from one of them. At a higher level, with friends and family, close proximity and detailed interaction may possibly raise them to the level of full consciousness, at least in my presence. Additionally, most of the world will not exist in any detail except as I move through it. What happens when I am not around would be some kind of superficial evolution of gross features almost like a soap opera on TV, with a basic script being worked through. This would be applied to both people I never meet and geography.

The key to escape is the realization that there is only limited computing power, and hence the world is going to be incompletely fabricated.

On one hand, this is as close as we get to being in a computer game running on seriously under-powered hardware, and on the other being in a dream. It is the latter which would appear to be most likely in terms of probability, given that this scenario probably stems from a single PostHuman mind. It may not even be a deliberate Sim. Even now, we constantly run simulations in our own heads, for example we imagine things like "what will happen if I turn up for work late?" by running through the scenario complete with models of the people involved acting as we expect they might. A PostHuman speculating, or remembering, what it was like to live in these times may well render a Sim to the level of detail we see in the world around us. In this case almost certainly the person running the Sim is myself after I have been augmented, and the likely time period is within this century.

The people I meet, the NPCs, are all aspects of my true self, the dreamer. As such they will be deeply connected to aspects of my own psyche. Indeed, as I consciously alter my psyche by adopting various persona or emphasizing aspects of it I will see this reflected in the world about me. This is the power of the solipsistic view of the world and has been recognized in many religious, and mystical traditions.

In which case we need to wake up to our true self.

One method is to retreat from the world hermit style. Ideally to a featureless cave and do as little as possible. The aim is to force all the computing power that is normally expended on the rest of the world to focus solely on me and my state of mind. If there is nothing for me to dream except the major character (me) then that character becomes all important and begins to merge with the higher levels of dreamer consciousness. The drawback is, naturally, that one gets to do nothing but meditate in seclusion.

Alternatively put a squeeze on the processing power available for the Sim, and hence force it to render the world in less detail that it might otherwise do. We do this by executing a series of programs that absorb a maximum of resources. In other words, we execute "people". Not by merely interacting with them in general in a robotic manner but in the level of detail that forces the Sim to raise them to full consciousness (or close to it). We get a group of people who are close to us and interact in as deep an intellectual and emotional basis as possible. Coupled with this the group should have a common aim or purpose. Such a group, and objective, only needs to be supported for a duration sufficient to bend a more malleable world to the only Will that now exists in any measure.

Or... raise the computational power required to maintain my own consciousness and that of my group by taking something like LSD. When people take LSD while having a brain scan it lights up like a Christmas tree, and subjectively time dilates enormously because so much is going on in the brain.

All recipes for a literal change of consciousness, but ask yourself – what happens to the characters in your dreams when you wake up.

Better hope you are not an NPC!

Scenario 6 – On The Edge

Again, this is a resource limited scenario where whatever machinery that runs our reality is close to the edge of maxing out. So, what would such a reality look like?

Well, for a start there would be various anomalies as the system failed to keep track of things or maintain logical consistency as computing requirements momentarily exceeded capacity. Yes, glitches is the matrix! You thought you put your keys on the table but they are not there. Except, when you look again, they are. You made a mistake. Or something did, and then rectified it. This is a world of one-off anomalies and unrepeatable synchronicities. It is a world where psychic phenomena (Psi) occur, but is not amenable to experiment because the system patches the holes as we discover (or create) them. Ghosts, UFOs, cryptids... As to how the patches are implemented, well at the more trivial layers of complexity it can probably be quite easily automated with deviations from the script fixed as they occur. Background helper programs would likely be used. In the original Unix operating system they had a name – Daemons. Most are stupid, but the ones handling more complex screw-ups would have to be high level AI.

Escape is unlikely, but there are possibilities as previously mentioned of locally overloading the Sim, which if pushed far enough might globally crash the system. It may well attract the attention of whoever is running it, and it would most certainly attract the attention of high level Daemons. However, the most likely outcome is that the system does a restore to a previous point. There was no nuclear war in 1985. If it is not a perfect reset there may be people with residual "false memories". They even have a name - "The Mandela Effect", so called because a disproportionate number of people of a certain age remembering Nelson Mandela died in prison and did not go on to lead South Africa.

Overall it is bad news, primarily because whatever is running it either cannot or will not provide sufficient power or programming consistency. It is exactly like the kind of thing we might do in around 50 or 60 years time when virtual reality and brain interfaces come of age to seamlessly(?) give us full sensory immersion. In which case we are either Humans on the edge of becoming PostHuman and are jacked in with an added memory block and false memory overlay, or we are AI programs ourselves. In fact, given the comments above about the LSD experience it does rather lean towards the latter. The best we can hope for, is that it is a version of San Junipero[56] as featured in the remarkable Black Mirror TV series episode of the same name.

So what other horrible possibilities are there? In no particular order...

I have taken a rather powerful designer drug that has dropped me into a lucid dream state, complete with psychedelic time distortions and amnesia. It ends... well, who knows?

I am a bored teenager circa 2100CE who is doing his history project on life 00 years ago – what was it like? (Complete with full pain settings). It ends when you die.

It's a game like "Life of Roy" in the TV series Rick and Morty[57] where a VR game encompasses and entire lifetime in a few minutes. It ends when you die.

We are hallucinations inside a Boltzmann Brain, which is a self-aware entity that arises from random fluctuations in a dead universe in thermal equilibrium. In which case there is absolutely no way out, or rather, any escape is to black infinity. It ends when you mercifully die in the perfect prison.

Conclusion

Most of the methods of breaking out of a simulation are probably going to happen anyway given time. Doing so is going to involve considerable risk and even if successful will probably not result in us literally "escaping", since I doubt that the next level up will provide suitable conditions for the life support of a Human body. More likely is that we will open up a communications channel between them and us, and *maybe* we will be given tools to exert a little more control over our reality.

Another thing to note is that anything capable of running "us" is by definition capable of running an advanced Artificial General Intelligence at least as smart as us and probably way beyond our individual capabilities. Do we really want to have a talk with the "caretaker", especially if we are messing things up?

Finally, as the Buddha might have noted, no matter what this reality may be, we can only know for sure the truth of two things – suffering and mathematics. Pain still hurts and logic is logic. Now... play the game.

Chapter 4

The Consciousness Process of the Simulation

Sean Byrne makes his living in the painting and decorating trades It's a job made for him as it allows his mind to wander where it will.He lives in Greenville Michigan with his wife Kathy and they have two adult children Mitchell and Julia who are his pride and joy.

There is a matrix created from models of reality subject to perspective and founded through mathematical conduction.

It creates through possibility and leverages from context to build beauty from mechanical systems. I wonder, what is its origin and purpose?

As I was driving the other day, a red-tailed hawk dove in front of my car and picked up a hapless chipmunk and carried it off to a tree to eat. I was thrilled to see this and yet mortified thinking about what was about to happen to that poor chipmunk. I couldn't help but wonder about the nature of a system that is founded upon such cruelty.

I sometimes experience what I call strong synchronicity of events and when I got home a video was posted on my timeline of another red-tailed hawk, this time capturing a pigeon and eating it while it was still alive in gory detail. Again, I felt a thrill and then total horror, but mainly I was again shocked into thinking about where it all comes from. Is there a system of logical control that has created such violence and judged it to be worthwhile and propagates it purposely?

I think that this is a reasonable question. There is lots of evidence to back such an idea.

Around our sun is an area called the heliosphere which is the distance that plasma from the sun is blown outward as a solar wind with enough force to encounter ultra-high energy cosmic rays from the deep galaxy. There is a "ribbon" of energy and particles there at the edge of our solar system that appears to be a boundary of the local interstellar magnetic field. This field of energetic particles may be only a small sign of the vast influence of the galactic magnetic field that extends well beyond our solar system.

Unknown until now, the direction of the galactic magnetic field may be a missing key to understanding how the heliosphere is shaped by the interstellar magnetic field and how it thereby helps shield us from dangerous incoming galactic cosmic rays. How magnetic fields of galaxies order and direct how galactic cosmic rays are resisted influences the environment of our entire solar system and our own environment here on Earth, including how that played into the evolution of life on our planet.

The Earth itself also shields us, protecting us from harmful radiation from the Sun and other stars. As its magnetosphere, the region surrounding the planet with its electrically charged particles, produces a magnetic field, it blocks solar and galactic radiation from scouring the planet. Without this field, life on earth could not exist.

Add to this the idea of a fine-tuned universe and it gets even more interesting. If you changed the force of gravity or the strength of electromagnetic or strong or weak nuclear forces the universe would be a far different place.

These are arguments for a universe with a creator.

To further this concept, we can explore one avenue of creation-simulation.

In 2003, Nick Bostrom proposed the simulation argument using what's called the *trilemma argument*.

> 1. "The fraction of human-level civilizations that reach a posthuman stage (that is, one capable of running high-fidelity ancestor simulations) is very close to zero", **or**
> 2. "The fraction of posthuman civilizations that are interested in running ancestor-simulations is very close to zero", **or**
> 3. "The fraction of all people with our kind of experiences that are living in a simulation is very close to one"

To anybody in our technologically advanced civilization the obvious answer is number 3 as it is easy to see that we already create simulated environments through books, television and now virtual reality.

As we continue down the path of ever increasing technological advance we will continue to create better and better simulated environments that are harder and harder to tell the difference from "real" environments.

Most simulation arguments use this fact to stipulate that it is very probable that we exist within a simulation already, as the odds of being in an original universe are very slight. The simulation hypothesis also accounts for peculiarities in quantum mechanics, particularly the measurement question whereby things only become defined when they are observed.

I see two possibilities. We are simulated which would entail a designer of some form, or we are generated.

To understand this, it is necessary to discuss Panpsychism.

Panpsychism is the idea that everything is conscious to some extent. To discuss this, I must state what I mean by consciousness. There are many ideas about what consciousness is- from a state of being to awareness of the environment. I feel that a good definition of consciousness should be quantitative or something that you can actively measure that also contains the axioms of the concept. The axioms I use are the abilities to understand the environment and to choose how you experience your path through that environment.

I define awareness as a reflection of the environment with meaningful decoding.

I define consciousness as an aware system that selects from its environment to create a pattern of continuity.

Next is understanding how consciousness forms and its place within the environment.

Nothingness is complete contextual uniformity. It can only be measured as a proportional statement such as if we achieve absolute zero within a proportion of the fabric of nature and then select for uniformity of substrate, we can measure conductance of contextual energies as they become increasingly superconductors across the space until nothing is available to measure.

Nature is the only reality. Everything measured is an abstraction created from definition derived through axiomatic metrics each relating to other metrics.

Any defined metric is incomplete and created from context which when altered in any part alters the whole fabric. This occurs instantly, but only from the perspective of the whole fabric and all abstractions are affected in relation to the degree of contextual similarity.

Any change in context creates contextual energy. Contextual energy is instant across the fabric but also measurable in strength by its nearness of context across the fabric.

The fabric is information. It is built from the algorithms of alterations of its context. All context is algorithmic in nature and all contextual energy is also a flow of information in algorithmic structures which must interact in a perceived manner to exist.

This means all reality is virtual in nature. These foundational algorithms are the protoconsciousness that creates nature in the state we perceive, though we are not the only scale that exists, but simply one reservoir of conscious perspective.

It is possible to use contextual energy to communicate information through scales using fractal energies of context such as the synchronicity of density as seen in Mandelbrot sets. The flows form distribution networks for consciousness models.

Information flowing within modalities of defined perspectives creates models capable of algorithmic selection processes which may begin to be used to give the models continuity of pattern so that they function either as mechanical aspects of some larger model or function on their own selecting from their environments to create their own patterns of continuity. It is in this way that time is created through the flow of continuity of patterns and explains why time is different from different perspective experiences.

This is protoconsciousness. A system selecting from the environment to create a pattern of continuity.

To do what we consider conscious activity also requires awareness, which reflects the environment with meaningful decoding of the information. We are consciously aware. We understand the environment we interact with and make conscious decisions to strive within that environment. It also takes a flow of information and a receiver of the information plus random chaotic induction to create a system truly capable of the ability to make a conscious choice- a conscious being.

This is a panpsychic universe. It is created from models of consciousness interacting at all scales.

A good example of this would be the human mind.

I consider a mind to be a virtual projection, generated by the brain with purpose of promoting continuity of the pattern of the self.

It is created by the entire brain, though thought and pattern recognition are speculated to be a product of the neocortex.

The neocortex is hierarchically organized into about 300 million pattern recognizers; each pattern recognizer consists of approximately one hundred thousand neurons in a vertical mini-column, and those pattern recognizers communicate with one another via a grid of hidden hierarchical Markov models, which are also each in a process of selecting from the environment as separate conscious entities forming societies which compete for dominance.

This also occurs throughout society as corporations and governments form entities that have their own voices and lobby for their own interests and people are just removable models influencing the greater pattern while competing as individuals with unique individual lives and interests.

With contextual energy, the context of any action is a large part of the future results of that action. If you step off a cliff the next thing you experience will be falling.

Contextual energy is created by all changing context. I experiment with it by asking: What is nothingness? When I view nothingness as contextual uniformity, it is a measurable thing that has unique properties as we go from say absolute zero to unifying the fabric of nature until we can no longer measure any attribute.

At this point within nature we have no context. There is no causality. You cannot predict the next state of the fabric. In a contextually deep concentration of the fabric, you will have causality in a contextually empty situation- you will have choice. The empty state of this contextual field allows for any change you wish to make to it. It is the blank piece of paper to your authorship.

Then we can go to the other extreme of density such as a black hole. When context is so crowded that we have nothing but an ultra-dense plasma that has created a warp so huge it seems capable of tearing the fabric, there can be neither change nor measurement of the unified context. Here, there is absolutely no choice nor any possibility of change while maintaining its density.

These holes in the walls of the fabric of nature are the ends of the universe. They may only be phase shifts that lead to other levels or fabrics or they may be examples of finite nature. What is obvious is that there is a progression of density of context and from that they are the bounds not just of the universe but also of conscious will.

Our minds are the way in which our consciousness is expressed. I like the question, *where is my mind*? Is it inside of the brain the way a movie is inside of a TV? I have no reason to think otherwise. It seems to me that our minds are generated by our brains in the same manner that an electric motor generates electricity by creating a movement of electrons using electromagnetic force.

The brain uses chemical electrical impulses to move information structures within the brain which could be protoconscious elements generating conscious fields of experience within a virtual space that allows the senses to put together a simulation of the environment so that the mechanical system of the physical body is able to prepare and plan its progression within its natural physical environment.

If we postulate that infinity is the only reality, everything defined is an abstraction created through the axioms which give it meaning. Axiomatic abstractions are models of reality subject to alteration of perception as perspective is changed through alteration of context. Context itself as it changes creates contextual energy which is the life blood and only true energy of the universe. We can now begin to see that consciousness changes from a primitive selection process to an advanced self-reflection of the universe, which as it gains understanding and depth in its creativity, does itself generate ever-increasing emergent qualities that expand the possibilities of self-control of experience for the consciousness modalities.

Evolution is a process by which the consciousness that is the universe self-selects its experience. All intelligence is the process of self-control of experience. It can be demonstrated that the process of evolution has been occurring exponentially since the beginning of our solar system and continues exponentially today as we become melded with our technology.

If all consciousness is algorithms, this may also mean that to be conscious, all you need to be is an algorithm of such complexity that you have the awareness and the ability to self-control yourself throughout your environment. We have seen many such algorithms worming through coded systems, bent upon destruction or manipulation of their victims, so it seems that the algorithms are growing and hunting. The simulation is creating the simulators. Now we see that it is possible that somewhere, sometime there are wizards using arcane spells of mathematical language, creating mystical creatures weaved into the fabric of space-time like cut and pasted photoshop game characters, technoshamans of post human simulacrums. Godlike creators of simulated realities who may or may not care about such little things as whether a pigeon feels horror as it is eaten alive. Seeing instead glory in its ability to strive and gamble with fate, not a tragedy, but instead the inevitability of living forces resisting the void.

What if the pigeon is a philosophical zombie? A construct that acts the part of a bird of prey so well that it only fools you into thinking that it has feelings? This could be likely in a simulated world.

How can I tell the level of conscious intelligence in another creature?

One day I was playing with my dog Savvy, I would say get him and step toward him and push him away and he would come back at me and pretend to try and bite me and I was stomping and laughing and we were both having a lot of fun till we got tired. I went back to working on my project working on a piece of 3-inch conduit when I noticed a little furry black spider with iridescent blue fangs doing exactly what Savvy had been doing. He was jumping up and down, dodging, and having fun like he wanted to play too. Using my fingers, I played with him in the same way saying "get 'em" and tapping by it while he came back pretending to attack but not quite touching me. Then again, we tired of the game and looked at each other. Then I slowly put my finger by him so that he could sense better who I am. He came carefully over and tentatively touched my finger with his mouth. Then he backed up lowered his back legs and looked up at me. I said you're a good boy and left him on the piece of conduit.

Looking back, I think that the energy between my dog and me was not a subjective thing but actual and palpable. The little spider (known as a Brave Jumping Spider) must have easily detected this energy and reacted to it. I may have misread the incident and though I think the likelihood of this is small, I could be mistaken. It is however interesting. I don't have the ability to map out the mathematical details of the exchange but I still feel that the communication was there. How is it possible for a spider to sense the game well enough to repeat it? I may not be able to measure the energy in its entire context but maybe the connection itself gives information. I know it happened and I also know it was a repeatable event.

I then ask what is the voice within the hydraulic system of pressure and energy of context that gives rise to the singularity of self from the myriad streams of parts. The voice is the expression of self-awareness. Pressure gives its process shape and direction. It is the pressure that gives voice its cohesiveness and it is the sense of direction derived from pressures that creates its sense of intention and therefore gives it meaning. And it is the mapping system of the mind that gives homeostasis to the process thereby defining the self through immune processes that define other.

And yet again we must ask what other has voice and so self. Does the wind truly have a voice or is it just making noise? Does the ocean have a voice? Must a voice be vocal or do other forms of energy carry self-contained awareness.

How about bees as individual or as a hive?

Once while at work painting dormers on a roof I needed to work around a lot of bees that had become very active in the early fall heat. There were about 6 dormers with windows and hundreds of wasps going about their daily business. I have gotten used to working with bees and wasps so I've learned that when I have to calmly yet insistently just go in and do my work using my paint brush to encourage the little guys to move out of the way so as not to get in the wet paint. As I've mentioned before I have noticed that insects can understand a lot of a situation by the energy of it so I like to tell the wasps, excuse me this is work that needs to be done and if they get angry I hold my brush up defensively and protest, hey I'm just a worker doing my job.

This was working well as usual until I was walking on the roof from one dormer to another when suddenly two large vicious looking wasps stopped and hovered about 1 foot in front of me halting me. One of them started to approach slowly toward me and I held out my hand toward it. I was in mind of the feeling of being pulled over by the police and questioned as to my identity and purpose. The wasp put his feet on my finger carefully and then bit me. I said alright that's enough and removed him from my hand with a little shake. He regrouped with his buddy and another bee shot quickly past from behind me touching my ear with a loud angry buzz. Remaining calm I said to the three wasps, look, I've cooperated with you and am going to continue to do my work. If you want to bring this further none of us are going to like it. It's your choice. I sensed confusion and then they moved off. Later I was working on a ladder painting a window with wasps all around me and was not bothered.

To me it seems obvious that the wasps had a voice and spoke very clearly to me. They questioned whether I was an outsider in their territory and what level of belligerence I was representing. Language can be defined as any system of formalized symbols, signs, sounds, gestures, or the like used or conceived as a means of communicating thought, emotion, etc.: the language of mathematics; sign language. The communication between us in this instance was an abstraction of a situation that needed to be processed in a manner that we both understood. The wasps had voice individually and yet had also acted in concert, representing the hive. I had a voice individually in the energies of my speech and actions as well as representing the very abstract idea of the group of humans who occupy the building.

It seems reasonable to assume that the voice is anything that we use as representation of our self and that anything that has abstract representation of self has voice and therefore to some degree consciousness regardless of its language. The important aspect is to delineate the embarkation of territory of the entity in question. A bee is one entity; a hive is an entity of a different scale. The definition of this territory is the definition of the being.

A wave in the ocean is not to my understanding a voice under this criterion. I will change my opinion if necessary should I discover evidence to the contrary.

I find the answer to my dilemma of the red-tailed hawk as a selector for my question of: is the universe a *top down* system of simulated realities given to us by a creator or is it a *bottom up* system that generates ever increasing emerging capabilities of better possibilities? Maybe it's even possible to have hybrid systems.
I will let you choose your own answer.

Chapter 5

The Simulation Sutra: Are hungry ghosts, poltergeists, Bigfoot, and UFOs a clue to the simulation theory?

Matt Swayne is an experienced science and technology writer, who turns complex information into compelling content. As a lecturer at Penn State college, as well as a Science and Research Information Officer he created and taught a course in science writing and communication. Matt has a vast expertise in science news writing, news writing, copy writing, blogging, and editing. His side hustle includes working with startups in AI, FinTech, Quantum Computing, and Venture Capital. He can be considered as a strong media and communication professional with a Communications degree focused in Journalism from Penn State.

Ever get the feeling you've been cheated?
-- Johnny Rotten, 1978

The Simulation Theory -- the notion that our reality is nothing more than, at best, a dream, and, at worst, a sham -- is certainly not a new concept. It's been causing awkward philosophical pauses and existential nightmares since Zhuangzi's butterfly dream and Plato stumbled across the fire burning in front of a cave and casting long, dark shadows into the recesses of our belief in a solid world. Although not new, recently the theory has moved from intro college philosophy classes and into mainstream consciousness through movies, like the Matrix franchise in 1999, and even modern philosophers, such as Nick Bostrom[51], who proposed his Simulation hypothesis -- Are You Living in a Simulation -- in 2003.

Finding out that the world we all took to be as solid as billiard balls whizzing across the green lawn of a local pub's pool table and as predictable as metronomic brass ticks of a grandfather clock causes angst among the theory's many critics. And there's another twist in the theory that gives us pause. If the universe is a simulation, who is the simulator- or simulators? Aliens, God, devils, a mad scientist, and even a video game-programming kid are among the lists of suspects. The list isn't very appealing to most.

In the following essay, I hope to suggest another possibility, perhaps the most unlikely possibility, that not only is the Simulation theory real, but also that evidence of both the theory's validity, as well as the simulation's originator can be found in phenomena that is typically tossed under the philosophical rug a little bit after childhood. But, with a few thousand years of Eastern thought, a deep dive into Western occult tradition, and even a few outlaw physicists and psychologists, you might find out that you are more intimate, more connected with the originator of this weird Maya -- or play -- than you might guess.

Current Theories of the Paranormal

The paranormal is a catchall term for a range of phenomena not explainable under current known laws of physics. The phenomena includes, but are not restricted to ghosts, demons, extraterrestrials, and creatures that are often referred to as cryptids, such as Bigfoot and the Loch Ness monster.

Scientists who accept materialism have the power of one commonly acceptable theory. It dictates that all phenomena can be reduced to matter and all matter is subject to natural laws. One of the difficulties for paranormal advocates is there is no common theory that can help explain the phenomena. In fact, sceptics and believers explain paranormal and supernatural phenomena, or the lack thereof, through a mishmash of physical and metaphysical theories. The most common and most accepted theory of ghosts held by believers and advocates is that when humans die, they become spirits. Some fail to transition to an afterlife realm for a variety of reasons. These spirits might, for example, not even know they died. The death might be extremely traumatic or violent. Some ghosts expect revenge. Others just like their houses and don't want to be displaced, thank you very much.

Nearly every culture has its version of these wandering souls with unfinished business. In some Asian cultures, for instance, ghosts and ancestral spirits, called "pretas" or hungry ghosts, haunt their old homes and stalk old battlefields. Similarly, they have unfinished business, but, in Eastern thought, hungry ghosts yield to more spiritual motivation, typically karma. Not unlike Dickens' Jacob Marley, the ethereal ties of hungry ghosts bind them to the mortal world so that they can burn off their considerable unresolved karma.

Despite this transition to a spiritual world, spirits often leave physical traces, according to paranormal theorists. Ghost hunters and paranormal investigators traipsing through haunted houses are often equipped with a range of scientific instruments to document these signs of physical presence as, paradoxically, proof of the spirit world. They may use electromagnetic field -- EMF -- detectors to record any changes in the field when ghosts transit from the spiritual to mundane planes. Digital recorders may pick up voices of these spirits, often referred to as electronic voice phenomena, or EVP.

Their work is often besmirched by cynics and critics, which isn't exactly fair because this group includes many well-meaning people. However, the critics do have a point. The use of equipment designed by materialist scientists to investigate a physical -- matter and energy -- world seems, philosophically, at least, paradoxical. And, the evidence is rarely taken as such by the materialist scientists. Also. Guess what? They never will accept it as evidence.

The same could be said about evidence collected by the investigators in the field of cryptozoology -- the field of research dedicated to finding creatures, like Bigfoot or the Loch Ness Monster, that, at least currently, exist only in mythology and folklore -- and UFO hunters who scan the skies looking for traces of alien spacecraft that can crisscross space and time, but have trouble touching down in a populated spot in broad daylight.

Super Natural Simulation

Most sceptics would end this chapter now. No evidence, a lack of scientific rigor, and a bunch of flimsy theories means the supernatural isn't so super at all. In fact, it's better labelled, the impossible, right? But, the volume of personal experiences -- subjective though they may be -- continue and, may even be, increasing in our society. People continue to see ghosts, patients experience near death experiences, and witnesses -- even highly trained fighter pilots -- watch UFOs dance across the sky.

In the literature of the simulation, these anomalous experiences -- let's be honest, ones that all of us have experienced at one time or another, whether they're minor synchronicities, feelings of déjà-vu, or large hairy ape-like creatures wandering around a campsite -- are typically described as "glitches in the Matrix." But they may be not glitches, but features of a much deeper and more provocative level of reality than a pop culture reference suggests.

Could it be that these definitely are glitches in the Matrix, but not bugs in some rogue AI-driven computer, but aspects of our own being? In fact, there are experts who suggest that paranormal phenomena may be evidence of a simulated reality, but not a simulation in the sense that it's a mechanical simulation, like a movie set, but that it's an organic simulation in the sense that it's a dream, one that arises from consciousness, your consciousness.

And just as a dream that arises in consciousness is imbued with meaning, along with the burrito that was way too big that you ate way too late last night, these weird phenomena can be read almost like messages from your deeper self, as, possibly everything else in your life.

Many Eastern religious groups and philosophical adherents have discovered this principle long ago.

The Buddhists' Diamond Sutra -- Vajracchedikā Prajñāpāramitā Sūtra -- considered one of the most influential of the Buddhist sutras ends like this:

So I say to you –
This is how to contemplate our conditioned existence in this fleeting
world:"
"Like a tiny drop of dew, or a bubble floating in a stream;
Like a flash of lightning in a summer cloud,
Or a flickering lamp, an illusion, a phantom, or a dream."
"So is all conditioned existence to be seen."
Thus spoke Buddha.

The world, then, is a ghost, according to the Buddha. It appeared out of we know not what and it is maintained by we know not how. The idea that reality is an illusion is not unknown in Western traditions, either. Closely related ideas, for example, are brought up even in a recent book by Jeffrey Kripal and Whitley Strieber in their recent collaboration on the book, Super Natural[58]. The book re-examines Strieber's famous abductions by -- what is commonly described as -- aliens. However, in Super Natural, the authors see the abduction phenomena, as well as a whole range of paranormal activity -- such as out-of-body experiences, levitation, ghostly visitations, mystical rapture, and more -- as symbols summoned by the real-unreal world of consciousness, or, perhaps, the extended Mind.

What I think this might suggest, then, in regards to a discussion on whether we are living in a simulation, is that the simulation theory might be more accurately described as an interpretation theory. In other words, we are not living in a simulation so much as we are interpreting the simulation from information in an infinite database.

In his book Authors of the Impossible: The Paranormal and the Sacred, Kripal[59] writes:

"That position comes down to this. The world is not simply composed of physical causes strung together in strictly materialistic and mechanical fashion requiring, say, a physics for their complete explanation. The world is also a series of meaningful signs requiring a hermeneutics for their decipherment. Whatever they are, UFOs "vibrate in phase" with our forms of consciousness and culture. We thus cannot even conceive of them outside or independent from their observation. This most basic of facts puts into serious doubt the adequacy of any traditional scientific method. Such methods, after all, work from an ideal of complete objectivity, which in turn demands an effort to eliminate all interference with the observer. But what if the observer is the very mode of the apparition? What if the observer is an integral part of the experiment?"

In other words, are we not the simulation itself? Or, is the inner working of the simulations not the inner workings of our selves, our minds? Again, in Mutant and Mystics[60], another of Kripal's books, the professor puts it more succinctly when he writes, "...we are haunting ourselves in the present from the past and the future via the ghost and the alien."

Conclusion

The Simulation Theory, as we've seen, is nothing new. In fact, the idea is ancient. It's existence as a theory is wrapped up in mankind's most basic existential question: Why am I here? Reality, the ultimate WTF, right?

The conclusion of this piece, unfortunately, won't necessarily help you answer that question. But, what anomalous phenomena does do, if you dare to stare at it, is point toward a deeper, more personal explanation for the existence of this simulation, or illusion, or whatever name you want to foist on this elaborate mystery. Even though this type of phenomena has worn the disguise of all the major aspects of "the Other" -- the devils, angels, gods, aliens, and ghosts that fill our mythologies and religions -- it actually tends to point away from those forces as things in themselves, but as representations. As Ralph Waldo Emerson says -- and I am probably taking way out of context -- "What terrible questions we are learning to ask! The former men believed in magic, by which temples, cities, and men were swallowed up, and all trace of them gone. We are coming on the secret of a magic which sweeps out of men's minds all vestige of theism and beliefs which they and their fathers held and were framed upon."

In the end, your search will go on. And your vocabulary will grow: Idealism, Nondualism, Multi-Solipsism, Monism, and Omnijective reality, are all ways religious adepts -- and now even some scientists --

are grappling with to describe the workings of the simulations, to somehow get their arms around this infinite net of Indra. But, as we know, words tend to point at the finger, not the moon. Or, in this case, a simulation of a finger, not the simulation of the moon.

Chapter 6

Parasitism and Perceptive Valuation and Presuppositions

Donald King grew up in Dayton Ohio. He was a foster child who eventually got adopted into a family of fifteen. As a young man he discovered he had a penchant for organized expression, which ultimately led him to pursue several paths of creative endeavor. To date he's a published musician, a writer and poet, a painter, and he has worked as a professional cook for many years.Somewhere in the area of ten years ago he had an experience that ultimately resulted in him gaining a perspective that allows him to see and process information and reality according to principles — that is, meta themes and grouping patterns that reverberate throughout reality. He tests his perspective constantly against science, academia and the very best critical minds he can find and get access to. Since his experience, he has built (and continues to build) a library of work that challenges many of the cornerstones and premises of and within science and academia; specifically in fields related to psychology, sociology, philosophy: ontology & epistemology, virology, maths, etc.

When first presented with the opportunity to contribute, I questioned whether my writing style, life experience, and general views on this subject would gel well with the tone and intention of the rest of the book.

When it comes to contemplating or exploring subjects, most times it seems the only real mental work required from me is that of searching for appropriate language and examples to use towards describing things I can actually see with almost physical precision and clarity. Exploring the strengths and weaknesses of concepts one's developing or scrutinizing is a different type of mental process than simply describing things according to how one sees them.

The latter thought process would generally apply to me and how I approach most subjects, including this one.

To some, the question "Is reality a simulation?" might seem almost laughable, because (ironically) I think that people's own experiences tend to indicate to them that reality is an emergent phenomenon comprised of events, cycles, and forms that appear structured enough to indicate systems of organization and interdependence, yet random enough to indicate a wide variety of dynamic systems of change. Often times, people's familiarity with their own experiences can cause them to equivocate their views and experiences with truth and reality. In this sense, things that challenge their senses of authority, specifically as it relates to their perceptions and preconceptions about truth and reality are generally met with disbelief or various forms of aggression.

I think perhaps the best approach to conveying my argument would be to create a closed system of reasoning, meaning that the ideal-types proffered in the reasoning system stand alone on their own merit, instead of relying on premises, views, or beliefs about this subject that were achieved from outside of my own reasoning ability. So besides an example I might use here or there, every concept I share in this chapter will represent a product of my own thinking. Thus this chapter itself should serve as an example of my argument-- by representing a closed system of conceptualization that functions mutually exclusive of any other theory or discovery pursued on this realm of contemplation.

Please note that the difference between an open system of reasoning and a closed system of reasoning is that an open system sort of serves to calibrate ideal types to or with the views and discourse on or related to a subject, by seeking support, edification and foundation from external perspectives prior to finding areas of contrast or disagreement, whereas a closed system of reasoning focuses squarely on the subject or topic itself, and in so, avoids impeding the conceptualization process with efforts to consider or adhere to authority. When and if possible, I prefer to avoid the conceptualization lag of authority-based reasoning. In other words: "You figure it out in your mind, I'll figure it out in mine and let's compare notes after the fact. Neither of us should have to wait to think, or learn how to format our thoughts so that we're building onto or necessarily working within the other's thought process or systems of organization and priority." Formatting one's thinking to fit with how others have thought before them necessarily handicaps their natural mind.

The question "Is reality a simulation?" actually breaks down into two distinct questions: Is reality as a whole a simulation; or is the phenomenon we experience and refer to as reality only a simulation that contains and/or was purposed for us?
To go about answering if reality as a whole is a simulation, let's establish some definitions by examining a few situational facts.

A simulation is necessarily a construct that was created with the express purpose of providing a controlled experience for those who are subjected to the experience, whether witting or willing. Something has to create not only the reality within the simulation, but also the physical, digital, and principle constructs that said reality would take place in. A simulation is necessarily the product of someone or something's intention. So if reality **as a whole** is really just a simulation, then a simulation supplied by what and who, purposed for what and who?

Something would have to Meta-exist independent of the perceived reality in order to create the construct for reality to take place in. Paradoxically, wherever that entity, group, or species resides is technically apart of (if not a more accurate depiction of) reality, indicating that reality itself cannot be a simulation. The possibility of reality being a simulation necessarily leads to a recursion problem.

This carries us to the next question: Is what we as humans recognize and experience as reality only a simulation?

Though this question is a bit more difficult to answer, my view is that what we experience as reality is not a simulation. The reason for this is because, as noted in my own work, elements of a system cannot perform above, beyond, or in spite of the systems they're born of, purposed for, and meant to help facilitate or contribute to.

Parasites cannot perform above, beyond or in spite of the parasitic mode and operating system any more than trees and other natural (autonomous) organisms can function above the respective ecospheres they're born of, purposed for, and meant to help contribute to.

That being said, parasitism (especially when manifest as collectives, in the forms of virus and disease) and the subsequent virion do function out of sync with systems and ecospheres they invade and co-opt, as their mode and subsequent propensities tend to wear on and destroy systems they are not authentically part of, yet cannot distinguish themselves from.

When you see things perform above, beyond, or in spite of systems, it indicates that they are not authentically part of the system(s) they're performing or being observed within.

In my own work, I refer to the mode of processing information and reality that humans perform through as "perceptive valuation". Long story short, perceptive valuation is a paradoxical mode of processing information in which the mind projects value(s) onto reality while simultaneously attempting to capture information/value from reality. This causes an irreconcilable conflict in the conceptualization process, wherein the mind becomes fundamentally unable to distinguish differences between what it projects onto reality and what it is actually taking in.

Perceptive valuation is a parasitic mode of processing information and reality. As I see it, perceptive valuation causes humans to think and behave like parasites. The effect humans have on the body of this planet is comparable to the effects any disease would have on any organism.

As I understand it, technology is a parasitic phenomenon, which means it cannot perform above or be achieved independently of parasitism and exploitation. Technology cannot perform above the system and mode of parasitism. Again, elements of a system cannot perform above the system itself.

Parasitism and what we commonly refer to as 'nature' represent two distinct branches of natural organization that spawn different types of cycles, life forms, systems of organization, and intelligences.

With perceptive valuation, there tends to be an over-reliance on projection, as manifest in the forms of things like meanings, biases, symbolisms, and beliefs — which are basically conceptual tools humans use to lend structure and navigability to reality and/or to make reality exploitable by means of. Because phenomena projected onto reality are highly subjective and skewed according to individual and group biases, the practice of attempting to determine what's "real" by extracting meanings from things that cannot be absolutely quantified, tends to be highly attractive to organisms that perform through this mode.

A few days ago, my good friend Antonin (summarizing the law of instrument) said: "To a hammer, everything looks like a nail."

According to my study of perceptive valuation, that statement should go something more like: "To a hammer, everything looks like it either wants to be or should want to become a hammer. In this sense, to some degree or another, every hammer believes all things are aspiring hammers. The hammer views itself and its way of solving problems as optimal. So the hammer "rightly" believes every single thing it strikes would choose to be a hammer if it could."

Or as my good friend Yanna put it, "The agent believer believes everything would choose to have its type of agency, if it could."

If put into the framing of the actual law of instrument (at least as captured in Abraham Kaplan's take on it), then instead of "Give a small boy a hammer, and he will find everything he encounters needs pounding.", I'd change it to "Give a small boy a hammer, and he will assume that all rational people should want to pound things the same way he does."

So in essence, the question "Is reality a simulation?" could be likened to an updated version of the 'tree falls in the forest' question, in the sense that humans can project infinity and all manners and scopes of possibility into perceptual blind spots, and then spend years contemplating and rationalizing the fictions they spin in their minds as being plausible.

Simulations are of interest to humans at our current juncture of intellectual growth and development as a species. They might seem relevant, perplexing, and complex to us now, as they are considerably newer additions to our technological capabilities and endeavors, but that doesn't mean they're necessarily interesting, useful, or impressive to species that (would) have already achieved and perfected them.

Ironically, I think this question serves better towards exposing weaknesses and propensities of human psychology than it does supply us with useful paths of conceptual investigation. Anthropomorphism, a phenomenon in which humans somehow elevate their view of themselves above reality (in thinking that reality somehow exists for them and their benefit) is at the root of the argument.
Another way anthropomorphism influences humans is by causing them to judge their view to be entirely consistent with reality (rather than 'above' reality), which is unlikely when the view spawned from the use of perceptive valuation.

In order for reality to be a simulation, someone or something would have to value human consciousness, modal thinking, and perceptual output enough to create an infrastructure that would facilitate and capture it. You could liken it to an evolved alien race, with non-compatible genitalia, harvesting human females for sexual gratification. It'd be like us thinking humans are so beautiful that other species must lust after human flesh in the same way that we do. It is wildly unrealistic to assume anything with the mental capacity to conceive of and create an artificial reality would be so impressed with the input-output processes of the human brain that it would construct an entire reality just to get our goods from us.

These scenarios humans present generally do one of two things — 1.

Either fails to account for preexisting circumstances and conditions that MUST precede the present state and condition of the experiential reality, or 2. necessarily elevate human perception above reality, so that reality is either something that emerges from the view(s), experiences, and imaginations of the species, or is something that is necessarily subject to and influenced by, not only their perception but by and according to their existence itself.

So, for humans in the modern era to question whether or not reality is a simulation indicates to me that we are fundamentally detached from reality and delusional because of our neglect of attention towards the sheer vastness of what we cannot conceive of from our limited perspective.

Human conceptions of reality are far too limited to even broach contemplating what and how reality is at large. There are so many layers of causal influence and systems of emergence, organization, orchestration, intention, and happenstance in play at any given time that attempting to assign any overarching theme or value to reality from our vantage point could be likened to a single (anthropomorphized) letter in Microsoft Office attempting to question if all of reality is simply a file (on the basis of it being a type of file in its own right, and then used to compose files, while belonging to a program that both functions and can be classified as a type of file that is also used to create files); all while paradoxically being ignorant to the physical computer its greater program is housed in and thus all of its knowledge and ability for conceptualization limited to, as well as being ignorant of the vast network of industries that transmute natural resources to create parts for laptop and desktop computers, the teams of engineers and software developers and designers that achieve the digital and physical infrastructures and systems that allow Microsoft Office to function as a computer program, in addition to the societies and (respective) economies that serve towards harnessing and disseminating energy and technological devices through service providers, all topped off by the years of innovation and refinement in technological endeavors leading up to whatever version of Microsoft Office the anthropomorphized letter would be experiencing reality through and/or by way of.

Simply put, too many things had to occur and converge in order for humans to even exist, and when considering the amalgamation of circumstances that have proceeded, facilitated, affected, shaped, and even forced the human state and condition to become what it is now, the sheer grandiosity of things we can't conceive of casts a shadow longer and wider than, ironically, we can even conceive of.

It seems my experience is somewhat unique, in that I don't really (or more accurately, am not necessarily required to) theorize about things. Instead, I simply explain the mechanical essence of concepts according to how I see them, and then how they fit and function together based on how they manifest within my perspective. An easy way to say that is (again), the only real mental work done when contemplating or exploring most subjects is that of searching for appropriate language and examples to use towards descriptions.

If the arguments I've presented in this chapter seem ignorant and dismissive of the greater discourse surrounding or pertaining to this topic, because they represent a closed system of thinking, then wouldn't that same principle apply to the subject of humans questioning whether or not reality is a simulation?

Or is my argument reality itself?

Chapter 7

The Intelligent Universe

Eva Déli is a Hungarian American with a background in cell biology and the visual arts. In cancer research she has coauthored more than fifteen peer reviewed publications and as an artist she had an active exhibition schedule for over a decade. Her creative energy has been shaped by a powerful visual comprehension combined with scientific rigor. She is a self-taught scholar of theoretical physics, neurology and evolutionary biology. Eva is also a keen observer of people. Her idea about the nature of consciousness is part of a cohesive, encompassing hypothesis on the natural world and evolution.

Abstract

In physics and evolution existing theories are challenged by a list of observed phenomena and experimental results. For example, the observed gravitational complexities in the universe could not have been produced by the existing matter content; time, horizon, inflation, cosmological constant and unification present unsolved and overwhelming problems. These pressing, irreconcilable conflicts and questions in contemporary physics can easily lead to a mentally powerless position, which inspire just-so answers. Just so answers short-circuit epistemological study and might have led to the simulation hypothesis that considers existence a computer simulation. However, many of the above problems find inherent solutions in a self-regulating universe. Entanglement leads to an evolution, which increases complexity, while maintaining low entropy. In biological systems, evolution engenders the highly intelligent mind, which is also a self-regulating system. Although they differ in their energy levels and organizational complexities, the elementary fermions, the mind and the universe form an interconnected, fractal structure and display self-regulating, closed energy structures that satisfy the principle of static time.

Introduction

Ancient philosophers noted that the universe's inherent beauty and harmony might originate from mathematical laws. However, in the age of computers, an observation of mathematical regularity in the physical world easily triggers associations with computers. The simulation hypothesis suggests that experience is a computer program, executed by an advanced civilization[51]. The simulation hypothesis sidesteps the irreconcilable conflicts and questions in contemporary science, such as evolution's ability to increase complexity, the direction of time, the interpretation of quantum mechanics, whether an encompassing theory of everything exists or the values of the fundamental physical constants. By kicking the existing problems in physics down to the next level, onto a new God, the simulation hypothesis expresses hopelessness about the nagging and divergent discrepancies in physics and discourages serous investigation. Thus, the simulation hypothesis is a diversion, which neither advances scientific discourse, nor encourages public trust in the sciences. As an advanced civilization, we must strive to solve the looming crises in the sciences, protect the environment and improve democracy. Finding intuitive, parsimonious scientific understanding inspires a practical world view, a la Occam's razor. The following is my ontological investigation for possible physicalist solutions to the current discrepancies in the sciences.

Discussion

The Principle of static time

According to general relativity, clocks and time can only be properly defined in the presence of matter. In other words, interaction is an essential ingredient of time. Without interaction, time stands still. The "problem of time" in physics is greatly unsettling, because the canonical quantization of general relativity by the Wheeler-DeWitt equation, which predicts a static universe, flies in the face of everyday evidence. Page and Wootters speculated that evolution of the constituents of the universe by quantum entanglement may produce a globally static system[61]. Evolution of entangled subsystems allows internal participants to experience correlations between a clock and the rest of the universe as time[62], while external observers are faced with a static universe. Entanglement acting through mirror symmetries and dualities of smooth horizon surfaces would always produce energy-conserving, symmetric spatial topology, which determines the state of the horizon or its energy level. Thus, the universe can fine-tune its parameters via self-regulation. Accordingly, interactions acting through entanglement would produce opposite entropic changes and would evolve toward polar singularities (Figure 1). The existence of poles, what we call white and black holes, is congruent with Einstein's field equations. While none of these poles is available for direct observation, black holes have been indirectly identified via their gravitational effects. Although highly resistant to observation via interaction, white holes should be the most energetic, negative-curvature and bulging fields in the universe.

Irreversible transformations, in which energy can turn into information or the reverse, naturally would lead to a static universe. Because entanglement creates antipodal curvature changes from a common point, it satisfies the Borsuk–Ulam Theorem (BUT). According to BUT, every continuous function from an n-sphere into Euclidean n-space maps some pair of antipodal points to the same point. Two points are called antipodal if they form opposite energy deviations, resulting in zero sums, from the origin. The Borsuk–Ulam Theorem was developed in mathematics and physics, but has been applied successfully in other fields. This leads to a self-regulation of the brain's electric activities, for example[63]. In elementary particles, Newton's Third Law dictates that force acting on a body generates a simultaneous and equal magnitude force in an opposite direction to the first action. The principle can be also applied to the stock market. News (even fake news) can disturb the stock price; but over the long term, the stock price tends toward its realistic value. In hungry animals, sensory stimulus triggers consumption, which recovers a relaxed, satiated state. The evolution of consciousness enhances the complexity of regulation, yet its basic premise remains unchanged. Stimulus gives rise to comprehension, movement, and memory. Although action and reaction are separated in time, they are always proportional. This is even expressed in the ancient wisdom of karma.

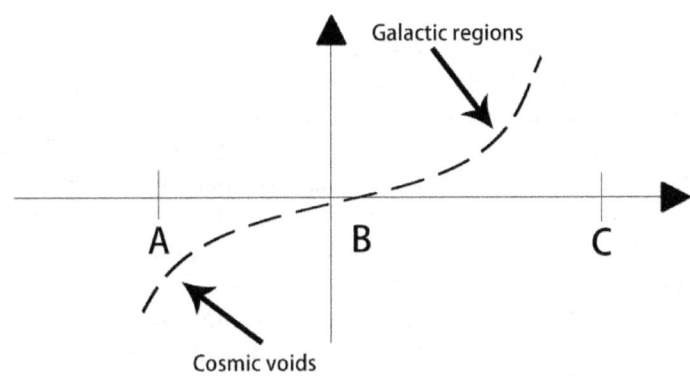

Figure 1. The Universe topology between its poles. (A) White hole with negative spatial field curvature; (C) Black hole with positive spatial curvature; (B) indicates Euclidean field curvature region. X coordinate is the age of the universe; Y coordinate is spatial contraction.

Landauer's principle

Landauer's Principle describes the convertibility of energy and information[64], which might be the foundation of interaction and the root cause of the incompatibility between general relativity and quantum mechanics. Originally developed for computation, Landauer's Principle was proven in increasingly sophisticated experiments[65,66]. The thermodynamic connection of energy and information means that erasing information dissipates a predictable amount of heat. In closed systems, Landauer's principle dictates that information enrichment generates temperature. Thus, information-saturated areas heat up, whereas energy-rich regions lose temperature. Our understanding of black holes supports this notion: their information-saturated horizon has high entropy and their immediate environment displays the highest activity and temperature. Furthermore, general relativity predicts that the infinite field curvature of black holes is a point-like singularity (devoid of space). Both statements can only be true, if black holes reduce onto two dimensions, which is congruent with the finding that black hole horizons are impassable firewalls and boundaries[67].

The irreversible operation of information erasure dissipates heat to the environment. The necessary amount of work is determined by uncertainty about the system — the more we know about the system, the less it costs to 'erase' it. This result suggests that 'information' might be linked to the concrete quantity, namely 'work', because recovering low entropy and low temperature causes the system to gain a finite amount of energy. In deep learning systems, the information input compresses time via Brownian motion, which increases entropy. Therefore, deep learning in closed systems might also occur via self-regulation, because information is transformed into energy due to entropic effects[68]. Moreover, it has been found that information transformation into energy slows down time[69], indicating a deep connection between entropy, information and time. The ability of self-regulating systems to maintain their low entropy state is a remarkable feat with regard to the second law of thermodynamics (The second law of thermodynamics dictates that the entropy of closed system increases over time. Here a local decrease in entropy is achieved by virtue of a global increase thereof). I propose to call the ability of closed, self-regulating systems to maintain low entropy, 'the low entropy principle.'

The Low-entropy principle

In the natural world broken, bent, misshapen items do not straighten themselves out; on the contrary, the physical world tends toward disorder, i.e., entropy increases. Civilized existence is a constant struggle for order. Creating or restoring order is the mark of organization and a sign of culture. Ancient landmarks, such as the Egyptian and Mayan pyramids, Stonehenge, or the Great Wall of China reflect the immense amount of time and effort invested in the creation of lasting order. In modern cities serious money is spent to maintain low entropy, i.e., law and order, by cleaning trash from public places. However, low entropy also means the ability to do useful work, which is possible only when objects differ in temperature, elevation, speed, etc. Temperature or potential differences equalize over time and form high entropy.

From ancient Greek philosophy to Immanuel Kant and Schelling, and in more modern times Ashby[70], Heinz von Förster[71], and others have considered self-organization related to thermodynamics[72] or as a philosophical concept. Schrödinger, who gained fame as the father of quantum mechanics, did suggest in 1944 that living organisms might have the ability to keep their entropy low. Recently, further development of his work in self-organizing systems led to the low entropy principle[73]. The ability of a system to maintain low entropy has immense and far-reaching implications. For example, in evolution self-organization is an ever-increasing complexity of matter into stars, planets, and living creatures. The evolution of the most complex system in the universe, the brain, has permitted social and economic progress.

The evolution of the cosmos

Einstein worked out his theory of general relativity according to the static cosmological model, in which the universe is "Stationary", i.e., both spatially and temporally infinite, and space is neither expanding, nor contracting. A steady state universe is isometric (the same from any observed direction), because it is Euclidean (Euclidean curvature is a field without any bumps or curvature in it, similar to a table top), and therefore shows no curvature; a low entropy state, which probably was true for the primordial cosmos. In the steady-state theory, the density of matter remains unchanged. However, while studying the movement of galaxies, Georges Lemaître and Edwin Hubble noticed that the universe is expanding. In 1998 it was further discovered that the expansion rate is accelerating. Because space, even cosmic voids, has a strict energy structure, expansion generates energy. This energy formation over cosmological scales contradicts one of the most important foundations of the stability of our world, the conservation of energy. Energy conservation is not just a physical law taught in schools, it is also our experience. Except for the obligatory loss of single socks in the wash, we always rely on the conservation of matter, such as our house, car and savings account. The same is true for energy, which dictates, for example, that a car does not work without some gas in the tank, cell phone batteries need to be charged, and pedaling a bicycle gives you good exercise. Such strict energy conservation is characteristic of Euclidean environments, such as our mild-gravity neighborhood. Energy conservation dictates that during interaction the volume of space remains constant. According to the principle of "static" time, interaction increases the differences in some cosmic indicators such as spatial curvature.

If we cut any small sample from our environment, string theory dictates that it would be made up of macro- and micro dimensions only[74]. Entanglement would shift the proportion of the dimensions in the daughter particles, as shown in Figure 2, but the global volume of the universe would remain constant. Over larger scales, accumulation of the micro dimensions would lead to contraction (and positive field curvature), whereas accumulation of macro dimensions would expand space (negatively curving space). The contrasting changes of the field curvature would form Polar Regions with opposite and complementary qualities[75], Figure 1. Therefore, the poles (white and black holes) represent the boundaries of the universe. It is known that black holes are information-saturated and have great field strength. As a consequence, both Einstein's field equations and the principle of static time tell us that white holes have zero information and zero field strength. The spatial field smoothly connects the opposing field curvatures of the poles and engenders low-entropy regions that are conducive for life. Therefore, the accelerating expansion of the universe might originate in the white holes, and be kept in check by the enormous field strength of the black holes (Figure 3).

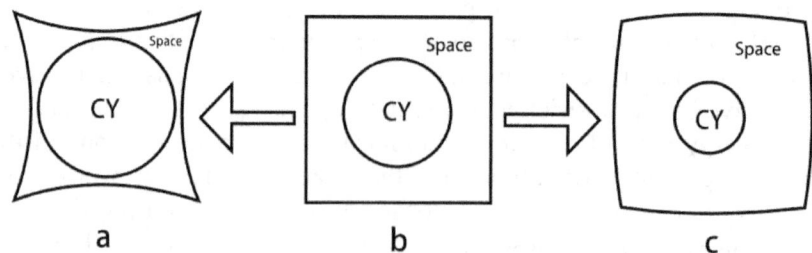

Figure 2. Particle entanglement (a)–(c) The micro dimension is indicated by the central circle, and the surrounding square represents the macro dimension;

(b) a sample of space; after entanglement: (a) macro dimension decreases in favor of micro dimension, causing space to contract; (c) in the entangled particle pair the micro dimension contract, whereas macro dimensional volume expands.

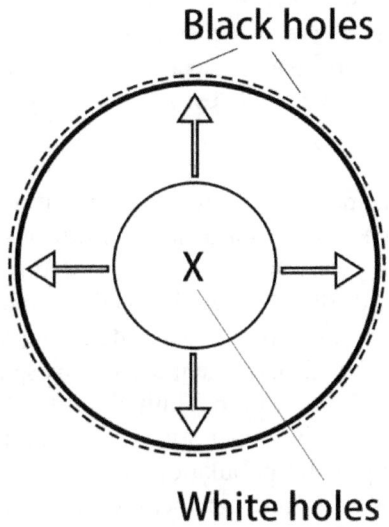

Figure 3. Structure of cosmos with white holes and black holes The expansion is generated by the white holes (indicated by white arrows). The outer boundary of space is formed by the contracting black holes; their great field strength stabilizes the universe and prevents runaway expansion.

Stages of evolutionary periods

The cycle of birth and death is a constant source of change and renewal in the living world. The dependence of biological systems on air, water, and nutrients can change sharply over time between excess and dire need, which permits the environment to regulate conditions for life. The limited availability of energy resources, such as nutrients, air, water and others at any one time leads to discrete evolutionary periods. Biological systems therefore have an expiration date, which turns evolution into a stepwise process[76]. Accordingly, evolutionary periods can be divided into three stages; energy excess, a stable equilibrium, and finally a broken energy-nutrient cycle. Because a perfect energy balance exists only in the Euclidean temporal curvature, middle section of the cycle, the low-entropy principle expresses the evolutionary period's optimal condition to support life. The beginning and end stages form complementary energy states: at the start of the period, nutrients are plentiful in the fresh ecosystem (reducing entropy) whereas, at the end of the period, disorder and waste accrue, which collapse the evolutionary period. Every repeated evolutionary cycle increases biological complexity.

Recent studies support the idea of a multistage evolutionary process (Figure 4). Important pieces of evolutionary innovations, such as a substantial part of the molecular architecture of the nervous or muscular systems, appear well ahead of their evolutionary importance[77]. In extreme environments, such as what occurs after major extinction events, or chaotic environmental and chemical changes, independent pieces of biological innovations can be pulled together into great biological complexity of adaptive phenotypes. Mutation frequency shows complex relationship with population number, because large populations remain more stable and can achieve greater fitness[78,79]. Moreover, in small populations mutations reduce entropy and subsequently entropy increases in parallel with the increase of population size, forming a U shape[80], as seen in Figure 4. Following environmental changes, the entropic effects of mutations regulate the emergence of new species[81] and formulate a stepwise evolution[82].
Social aspects of biological systems also support the idea of multistage evolution. The so-called prisoner's dilemma examines cooperation between two completely rational individuals. During the first vibrant, energetic stage of evolution, species from bacteria to fish to humans appear to lean toward generosity, leading to cooperation and altruism. The generosity appears most prevalent when mutations occur at an appreciable rate, which is often true for the first stage of evolution! Although the best interest for participants is to cooperate, evolutionary studies show that defections in the population gradually reach a tipping point, after which generosity and cooperation disappears[83]. In the third stage of evolution, cheating becomes the only feasible choice. The above findings show the field's control over individual behavior, as

recognized by Harsanyi's[84] pioneering work in game theory. Game theory shows that one person's gains result in losses for the other participants. Because the total energy supply of the ecosystem at any one time has an upper limit, the total history of the evolutionary period forms a zero-sum game.

The brain has been known to have immense energy use, by far the highest of any organ. Although only a tiny, holographic fraction of the total sensory experience reaches the cortex, most of the brain's energy use is applied for the maintenance of sensory readiness. As mentioned earlier, knowledge about a physical system changes the cost of information erasure. Likewise, the information value of stimulus depends on past experience, the brain's alertness and other personal factors. Because sensory processing is automatic and involuntary, sensory stimulus can powerfully activate electric oscillations in the brain. Sensory processing and their mental consequences are compulsory and question the existence of free will. Furthermore, comprehension and interpretation of visual, auditory, and tactile experiences require energy due to enhanced brain frequencies and trigger the consumption of complex organic compounds, which reduces the environmental entropy. Instead of the brain having control over the environment, life is one important piece in the environment's complex entropic regulation.

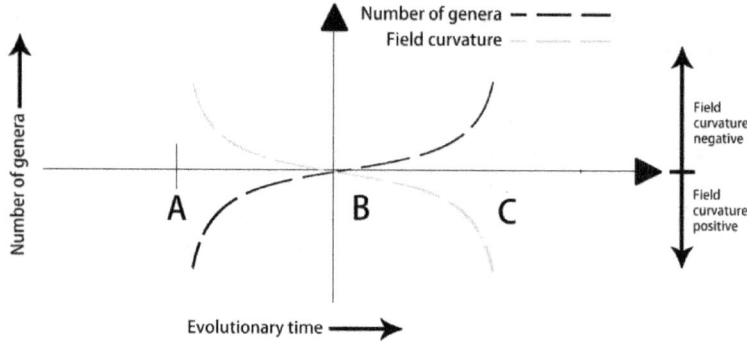

Figure 4. Changing entropy and biodiversity during an evolutionary period The evolutionary period can be divided into three stages A, B, and C: (A) has negative field curvature. The decreasing entropy results in rapid, seemingly arbitrary evolutionary innovations and jumps, which introduces new species increases the population number. The Euclidean region, which represents the stable, low entropy stage, characterizes the second stage B; and in the third, final stage C, genetic refinements lead to morphological differences. Positive field curvature corresponds to increasing disorder, which ends the evolutionary era. Biodiversity increases in the first and third stages of the evolutionary period. However, in the first stage, decreasing entropy allows evolutionary innovations to emerge by the reorganization of genetic material, whereas in the third stage genetic changes spread in the ecosystem and increase entropy.

Conclusions

Conceptual leaps are rare in the sciences. When newly developed techniques accumulate conflicting data with existing theories, daring and even bizarre solutions are proposed in an attempt to keep the existing paradigm. For example, in the nineteenth century, ether was invented to explain the spread of gravity. Thus, great schisms in the sciences necessitate radical ideas that change the accepted vision of the field. Accepted theories, such as general relativity became a magnet for scientists; countless applications explore and develop the theory's potential, the evolving rules reduce its flexibility. Any emerging contradictions are patched up by increasingly fantastical explanations and today this might have inspired the simulation hypothesis.

The principle of static time and string theory show that entanglement formulates a cosmic evolution which gives rise to polar singularities, called black and white holes. Nevertheless, the high entropy poles maintain low entropy regions, which great complexity and evolutionary potential, between them. Other recent hypotheses, such as the principle of static time, or the black hole firewall hypothesis are well-proven, yet remain unutilized in mainstream science. Considering the universe as a self-regulating, coherent system that engenders its own evolution toward increasing biological complexity. In self-regulating biological systems Maxwell's Demon utilizes the second law of thermodynamics to maintain low entropy and produce complexity, organization and intellect. In this way the mind is part of a global organization. Applying Landauer's Principle for the brain shows how energy/information exchange increases synaptic complexity via comprehension, memory, movement, and other reactions. The universe's three energy levels, the elementary particles, the mind and the universe, form a fractal system. Therefore, the universe's evolution and complexity can be explained within a physicalist framework, which satisfies Occam Razor.

Chapter 8

Pandeism and Simulation theory

Knujon Mapson is a student of the revolutionary evolutionary theological theory of Pandeism, a constant contributor to various discussion fora on the topic, and an occasional coordinator of discussions amongst other pandeistic thinkers. Knujon co-authored and edited the successful masterpiece on the topic of Pandeism called: "Pandeism, An Anthology".

The first question which confronts us in making a complete rational examination of existence is whether we can truly know that we exist at all. There are, after all, an infinite number of conceivable circumstances wherein we think we do (or one of us does; or *something* does), and yet we do not. Two potential reasonable possibilities, not necessarily mutually exclusive with one another, are simulation theory (which is essentially a scientific proposition) and the theological theory of Pandeism. The gravamen of simulation theory is that what we perceive as our existence is a consequence of some entity creating a program which simulates a Universe, and running this program on a sufficiently powerful computer, perhaps so as to see what happens. The gravamen of Pandeism is that our existence is the consequence of a being comparable to a theological deity wholly becoming our Universe, perhaps again so as to see what happens, but from a more experiential vantage point.

Either way, if our Universe is a simulation, then it is by definition a "created" thing. And so it must be a simulation with a reason. It is possible, one might suppose, to imagine some alien entity or race of immense intellect creating random simulations for no purpose whatsoever, but that isn't an especially rational view. More likely, the simulation—or comparable creation—serves some purpose, and most likely, that purpose is one which is primarily to the benefit of whoever set forth the simulation in the first instance.

A World of Possibilities:

Naturally, there exist many other possibilities along these lines. We could be unintended manifestations of the imagination of a higher power with no real substance; though if this is so, it would require at least this higher power to exist, for some reason. You, the reader (or I, the writer), could be a brain floating in a jar as part of somebody's experiment; or either or both of us could be a computer program made to think we are an independent being, or to think we are a brain floating in a jar imagining itself to be an independent being. One of us, between you and I, may be the only thing of our kind which exists, or the only thing which exists at all. Our eyes and ears, even our sense of touch, may not be trusted, for as 'real' as they seem to us, they are only (even in our "reality") electronic signals being fed into our brain for processing. Signals may be replicated, falsified — and that's simply assuming that what we believe as to electricity and biochemistry is the truth, and not itself an illusion.

This question has bothered many before, and will rightly continue capturing the interest of many modernly and in the future. René Descartes, probably best known for the pronouncement '*Cogito, ergo sum*' -- 'I think, therefore I am' -- more fully had actually argued "*Dubito, ergo cogito, ergo sum*" (I doubt, therefore I think, therefore I am)[85],[ii]. The core of this argument is that each of us has an existence which is at least sufficiently certain to result in our belief that we exist (or at least, you, the reader of these words believe that you exist, even if the rest of us may be illusions). Even if our existence is nothing more than an illusion which is good enough to fool us, at least we have 'sufficient' existence to be fooled by it[iii].

---

ii In actuality, Descartes being French, his initial phrasing of the statement was 'Je pense donc je suis.'

iii Naturally, there is some awkwardness is speaking of having a 'sufficient existence' for any purpose—after all, how could one exist non-sufficiently? Logically one either exists or one does not exist. If the reader will pardon my extravagance, you may substitute 'sufficient awareness' or 'sufficient

This enunciation was actually not the first attempt to forge such a connection. Karen Armstrong writes that "anticipating Descartes, Augustine argues that knowledge of ourselves is the bedrock of all other certainty. Even our experience of doubt makes us conscious of ourselves." Augustine's formula was 'Si [...] fallor, sum'; 'If I am *mistaken*, I am.'[86] A variation of this form was enunciated in the Eleventh Century, by Abū Alī ibn Sīnā, a Muslim scientist and philosopher better known in the Western World as Avicenna. While imprisoned in a fortress by the emir of Hamadhan, Avicenna penned the thought experiment of the 'Floating Man.' Avicenna implored the reader to imagine himself afloat in the air, cut off from all senses, even those of inhabiting his own body. Because such a person would still have self-consciousness, wrote Avicenna, this self-consciousness need not reflect any physical thing; instead, it must be a substance all its own.

Descartes follows the path to confirmation-of-the-self blazed by Augustine and Avicenna. The particular application of doubt in this formula was actually Descartes' great innovation — even if our sole consideration is to doubt our existence, we must have existence at least sufficient to allow entertainment of such doubt:

> But I have convinced myself that there is absolutely nothing in the world, no sky, no earth, no minds, no bodies. Does it now follow that I too do not exist? No. If I convinced myself of something... then I certainly existed. But there is a deceiver of supreme power and cunning who is deliberately and constantly deceiving me. In that case I too undoubtedly exist, if he is deceiving me; and let him deceive me as much as he can, he will never bring it about that I am nothing so long as I think that I am something.[87]

---

consciousness,' though confessedly something can exist without being either aware or conscious.

Descartes finds his own existence so confirmed, concluding "after considering everything very thoroughly... the proposition, I am, I exist, is necessarily true whenever it is put forward by me or conceived in my mind."[87] But he proceeds from this promising start to employ variations of the Ontological Argument—the proposition that a conceivable best-of-all-things is even better if it actually exists, and so, must exist—to posit a most benevolent God. Others have convincingly argued that the Ontological Argument can truly relay nothing as to a subjective characteristic such as benevolence, or even to existence as a quality which renders an existent thing better than a purely conceived thing. But it is this which Descartes uses as a basis to believe that the world he perceives is not illusory, as such a God would not allow him to be fooled into believing in an illusory world.

The conclusion reached by Descartes is, unfortunately for his advocates, undone in the light of modern technology. In at least two ways, our capabilities in this regard are breaching the barrier to humans indeed being able to engage in precisely the deceits which Descartes discards as prohibited by a good God. Firstly, we will soon be able to create computer programs which simulate a mind, and feed these programs artificial indicia of sensation, possibly even fooling them into thinking they are persons experiencing the world in the way which we are used to experiencing our world. Secondly, we have already created simulations which humans can step into and engage with, with little suspension of disbelief, and will in much the same time frame achieve the ability to create artificial stimuli which are sufficiently real to fool actual humans.

Naturally, these possibilities have been expounded upon in fiction. The latter is in actuality the core concept of the *Matrix* movie series, wherein the protagonist learns that the experiences he and other humans share in the world they believe to be real are in actuality illusions created by a computer, their real bodies being housed in an enormous network of pods. The former has been depicted in various episodes of the later incarnations of the *Star Trek* television series, wherein the 'real' characters engage in recreation on 'holodecks,' simulated environments complete with computer-generated people who seem self-aware, but may be unaware that they are simulacra. These holodeck people may simply mimic self-awareness without actually experiencing it. But two episodes relate a certain instance of artificial self-awareness. In one episode, 'Elementary, Dear Data,' crew members spur the creation of a holodeck program including a self-aware simulation of the character of Moriarty, a recurring villain from the Sherlock Holmes stories of Sir Arthur Conan Doyle. This character—accidentally designed to be an intellectual challenge to the android character Data, instead of to original protagonist Holmes—seizes control of the Enterprise, relenting when it is revealed to him that he is nothing more than a simulation, unable to leave the holodeck. In the later episode, 'Ship in a Bottle,' Moriarty attempts to force crew members to find a way to make him 'real,' but is instead tricked into simply entering an alternate simulation—a self-aware artificial mind, tricked (once again) into believing that it is experiencing a real existence in what is actually an artificial environment.

Fictional accounts aside, when such technology does exist in the relatively near future, it will inevitably actually be used on some people, either with the intent of deceiving them, or at their request, for purposes of entertainment. It is entirely possible that this technology will enable the user to suppress their 'true' memories of the world, providing them with an existence which they subjectively believe to be the entirety of their experience. There is, furthermore, no reason at all why the recipients of these illusions ought to be constrained to experience only the sorts of sensations which accord with our familiar laws of physics. And so, new technology could even lead us to experience sensations which are unjustified by anything which is or could be in the real world. Such illusions are perhaps not even confined to a technological future. They exist still today in the phenomenon of schizophrenia. People experiencing what we refer to as insane delusions are divorced from the hard-nosed physical reality which surrounds them. They see and hear things which are simply not there with as ready a facility as any machinery might induce.

The World Is Already Illusory:

Technology and insanity aside, the understanding of the world expressed by the sanest and most rational amongst us deviates far from an underlying reality of which we ought, in all good conscience, to be aware. For example, the appearance and consequent belief that the world is *stationary* are illusions. As one of the earliest Greek philosophers, Heraclitus of Ephesus, observed, the world — and we too — are in constant flux, forever changing though we may have the *illusion* of homeostasis. In the last century, English astrophysicist Arthur Stanley Eddington once set forth that a floorboard plank on which he might step "has no solidity or substance; to step on it is like stepping on a swarm of flies."[88] He spoke, naturally, of the atomic nature of matter, roughly expressing that the matter which a floor is made of is, upon very close examination, discernible as a swarm of subatomic particles constantly flying up against the corresponding swarm which is our feet; which in turn creates the illusion, very convincing to our macroscopically inclined powers of perception, of a solid floor.

A decade later, theologian Whately Carington resoundingly criticized Eddington for this analogy, exclaiming that it was "not only nonsense, philosophically, but remarkably bad science, for if the plank is a swarm of flies, Sir Arthur Eddington is one also."[89] Doubtless Eddington himself would immediately agree that, yes it is a swarm of atomic particles which makes up our own bodies, and if the floor be flies, then so be he. Physicist-cum-theologian Paul Davies observes in dismissing the famous comparison of a wristwatch to a clockwork Universe, "no analogical argument can amount to a proof; the best it can do is to offer support for a hypothesis… [T]he degree of support will depend on how persuasive you find the analogy to be."[90] As difficult as it is to analogize from our everyday surroundings to the subatomic or galactic levels, it is fairly impossible to compare any mundane thing which exists to the power behind the creation of our Universe, no matter whether or not this power is 'infinite.' But it is harder even still to illustrate certain abstract concepts without asking the reader to visualize something more familiar. But Carington's attack nonetheless carries its intended emotional weight; we reject at a gut level the idea that we are made up of "a swarm of flies," and incredibly that rejection is sufficiently strong to make some reject the analogous proposition that we (like the floor we stand upon), are made of a swarm of atoms. The state of flux was part of what Eddington was trying to convey in comparing the electrons in the floorboards to a swarm of flies, and yet it is Eddington who the critic seeks to make seem less rational for pointing out a scientific truth!!

After the death of Swiss engineer Michele Besso in 1955, Albert Einstein—a close friend of the engineer—wrote a letter of consolation to Besso's family, which included what would become one of Einstein's most famous quotes. "Now he has departed from this strange world a little ahead of me. That means nothing," wrote Einstein, explaining that "people like us, who believe in physics, know that the distinction between past, present, and future is only a stubbornly persistent illusion."[91,iv]

We, you see, are observers of a limited spectrum of wavelengths of light, of limited scales of size and of time, and very probably of limited dimensions of existence itself. Imagine you have a piece of paper laying on your table. Just a regular sheet of blank notebook paper. Now imagine an ink stain appears and begins slowly spreading from the center of the paper, towards the edges. This is, naturally, a two-dimensional view of expansion. But now, imagine that you take the paper and roll it just until the edges touch, and then you tape those edges together. Now, what do you have? A tube. A cylinder. And the ink stain spreading towards the edges of the paper is going to run into itself when it hits those edges. If you chart out all the wavelengths of light from shortest to longest, a little band of color left of center will represent all the wavelengths of light we actually see, against the backdrop of all the wavelengths of light there actually are.

A similar phenomenon occurs with our capacity to observe scales. We can "see" things in a narrow band of possible scales, ranging from a speck of dust to perhaps an entire city viewed from a high vantage point. Now this may seem like an enormous range, but it is actually a small portion of all possible scales. And the more we focus on seeing small things like a dust speck, the less we are able to see their context, while the more we focus on large things like a cityscape, the less we are

---

[iv] Einstein was prescient about Besso's death being "a little ahead" of his own—which followed barely over a month after.

able to see minute detail. And there is an entire universe of ranges of things that are too tiny to see even with the most focused eye, and just as many things that are too big to see.

And, yes, we can model every particle in the swarm of subatomic particles and take pictures of entire galaxies, but it is nearly impossible for our minds to wrap around the actual size and relative behavior of objects at or below the cellular scale, or at or above the planetary scale. The same limitation applies to dimensionality/ We perceive three or four dimensions, depending on whether the model considers "time" as a dimension, but physicists have proposed there may be many more (it is common to hear talk of ten-dimensional space in superstring theory, eleven-dimensional space in M-theory, and in bosonic string theory a whopping twenty-six dimensions!!)

So, going back to the ink spot spreading on the surface of the now-tubular paper, what we perceive as a Universe spreading in three directions may actually be as rapidly contracting to a collision with itself along a fourth (or fifth, or eleventh, or twenty-sixth) direction outside of our capacity to perceive dimensionality. All of this relates to the question, how may we be sure that anything we perceive is real? Since the solution proposed by Descartes is undone by technology, the unsettling truth is that it is simply not possible to be absolutely certain of *anything*. Bertrand Russell may have put it best when he declared: "To teach how to live without certainty, and yet without being paralyzed by hesitation, is perhaps the chief thing that philosophy, in our age, can still do for those who study it."[92] But is the illusion of some semblance of philosophical certainty meriting pursuit to the *n*th degree — or are there not, indeed, more important questions? Thus, perhaps we may instead only determine whether it is *reasonable* to believe in the reality of a particular proposition, and what probability to assign to such proposition.

A World Illusory Even to God:

Another of the earliest Greek philosophers, Xenophanes of Colophon, openly disdained the anthropomorphic nature of the gods adored by the Greeks, writing as to how:

mortals suppose that the gods are born (as they themselves are) / and that they wear man's clothing and have human voice and body / but if cattle or lions had hands / so as to paint with their hands and produce works of art as men do / they would paint their gods and give them bodies in form like their own/ horses like horses, cattle like cattle. [93]

In another fragment, Xenophanes carried this comparison to the different races of men:

Greeks suppose the gods to be like men in their passions as well as in their forms / and accordingly represent them / each race in forms like their own/ in the words of Xenophanes / Ethiopians make their gods black and snub-nosed / Thracians red-haired and with blue eyes / so also they conceive the spirits of the gods to be like themselves.[94]

But are we not now the same in our techno-anthropomorphizing? Modernly, we occupy a world of computer programs and programmers who write them, so perhaps it becomes more and more natural for us to suppose our gods to be akin to programmers as well. And flowing from that supposition, we ourselves must be within the program, or we are the program itself. Incidentally, German physicist and theologian Max Bernhard Weinstein wrote of Xenophanes that the latter spoke as a Pandeist in stating that there was one god which "abideth ever in the selfsame place, moving not at all" and yet "sees all over, thinks all over, and hears all over." [95,v] And is that not how a computer running a simulation would seem, as to the simulation running within it?

---

[v] p.231 of ref. 93: "Pandeistisch ist, wenn der Eleate Xenophanes (aus Kolophon um 580-492 v. Chr.) von Gott gesagt haben soll: "Er ist ganz und gar Geist und Gedanke und ewig", "er sieht ganz und gar, er denkt ganz und gar, er hört ganz und gar."

There is a very English book, charming and whimsical in its set-up, but still about geometry and its implications, called *Flatland: A Romance in Many Dimensions*. Late in the book, the reader is introduced to Pointland. The story has already by this time visited both Flatland and Lineland, where the imagination of the author becomes increasingly teased out in explaining how 2 and 1 dimensional characters can have societies. But in Pointland, we have no society, because we have only one individual: The King of Pointland. Stuck in a universe which is confined to one point, himself, he imagines that he is everything, and that everything is him. And so he becomes his own God, with nothing to do but spend all his time soliloquizing about his own grandeur and importance. The appearance of the King of Pointland seems an existentialist statement that was both anticipatory and beyond the scope of English philosophy. The narrator, the two dimensional "A. Square," and his companion, the Sphere, look down upon the King of Pointland and ask if he can be helped. They even yell to him, but he assumes it is his own voice that he hears.

After the attempted rousing of the King of Pointland from his dimensionless thinking, this King returns to an eternity of contemplating his own wholeness, unaware that he is the smallest thing in the universe. Bur what if we are *all* the King of Pointland? The question of what it means for something to be "reality" or a "simulation" is of infinite consequence here. After all, if we perceive something as real, and our capacity for perceiving is interwoven with the result of perceiving, then whatever is perceived is "real" relative to the perceiver.

There is, then, one proposition which no deity possessed of godlike powers would ever be able to know the truth of, and that is whether it itself is simply a construct within an even greater reality beyond its perception. This may sound odd and impossible, but it is aptly demonstrated by Gödel's Second Incompleteness Theorem, which proves that no system can be fully "solved" from within the parameters of the system itself.

Imagine a point in space. Now imagine that a straight line begins at that point and extends infinitely onward in one direction from that point.

Flat space, please, we don't want that line curving around and becoming itself. Now imagine a second point right next to the first, with a parallel infinite line extending infinitely outward in one direction. So now you've got infinite nonintersecting lines. If either line could be aware of its own infinitude, it might imagine itself to be the only thing in existence capable of being infinite. Now, let us take the second infinite line and move its starting point backwards an inch relative to the other line. Now you have two infinite lines, but one is an inch longer than the other.

Now let's take that longer line and stretch it along a second dimension, making it a flat plane with a few inches of height (and still infinite length). We keep it parallel to our first line, so they still never intersect; the first line still has every reason to believe itself to be uniquely infinite, even though it is right next to a thing that is not only an inch longer, but has an entire additional dimension of substance. And now we will take the second object (which is now a plane) and curve it into a cylinder which completely surrounds the first line, but continues never touching it. And just for the sake of it, we'll take that end of the cylinder that goes past the first point and curve its edges toward one another, into a sort of hemispherical cap which closes off that end of the cylinder (still without touching the original infinite line). Indeed we could do all of these operations to the first line and make into an infinitely long tube with one end beginning with a hemisphere, and still have it completely contained within a slightly larger tube with which it never intersects.

Now you might at this point be thinking, 'yes but the gods envisioned by human faiths are not *mathematical* constructs, they are not tubes; so what's the point?' Well, you would be correct that man-described deities tend not to be mathematical constructs, but human ideas of the ultimate are necessarily constructs nonetheless, for they are invariably described in human languages, and with human words and concepts such as 'infinite' and 'absolute' (and really those are concepts which, if not outright mathematical, are subject to mathematical discussion). Lest we delve too close to describing perfectly spherical gods in a vacuum, let us get straight to the heart of the capacities philosophically ascribed to deity-models. Able to do anything; possessed of all knowledge. But wait, there's the catch. No matter how much knowledge a god possesses, it is impossible for it to know that it is not missing some knowledge of which it is unaware, such as the existence of a greater surrounding entity which contains it.

Remember Neo, in *The Matrix*; once the nature of the Matrix was revealed to him, he found that while outside of it he was a normal human, but inside of it he had, essentially, superpowers. But how could Neo outside the Matrix know he was really *outside* the Matrix? After all, though things smelled and tasted and felt different, all of that could simply be another trick of the computer, sending signals which let Neo to think that he was experiencing those differences—Neo's entry into the "real world" would be no more real of a transition than the Moriarty of the Star Trek holodeck entering the "real world"—an artifice contrived by an external power greater than its experiencer. Naturally, the lack of superpowers in the 'real' world (and existence of superpowers in the 'false' world) would be trivially easy to conjure up through the signals sent to Neo's brain (if he has a 'brain' at all, for Neo's entire existence could be as a subroutine within a larger program, one programmed to believe itself to be an independent organic entity). And no amount of contemplation would then allow him to peel back the falsity of the 'real' world outside the one he already knew to be false.

But suppose instead of being programmed to believe himself to be a superpowered human, Neo was programmed to believe himself to be an omnipotent/omniscient deity? In this program, anything which he wished to bring about would instantly present itself to him as having transpired; any knowledge he wished to have would instantly manifest itself in his mind (or, at least, a convincing simulacra of such knowledge would). If he wished to set forth living planets orbiting burning stars to fill hundreds of trillions of galaxies, and know the every movement past and future of every atom in them, he could (if sustained within a powerful enough system) do so with a thought. But none of this would prove that it was Neo's inherent godliness causing such things to come about, for it could always be an even greater being, an imperceptibly nonintersecting surrounding cylinder of a being, providing all of these experiences.

Even if in some relative sense the Bible or the Qu'ran or the Bhaghavad Gita were true (such that there was a Creator entity which set forth all that we humans are able to perceive at least, and all the events recounted transpired as set forth there), even this Creator entity could never know that everything it was doing was not part of some infinitely greater Creator-entity's thought experiment. (Here is an inverse proof of this -- imagine a true all-powerful being not sustained within anything greater than itself; such a being could, by definition, create a slightly lesser being which believed itself to be the true 'only all-powerful being,' and which would by dint of the true all-powerful being's all-powerfulness, be absolutely unable to detect the greater being of which it was part).

And if no hypothetical Creator of our entire Universe could possibly actually know whether it was the ultimate being, then surely we spatially and temporally and intellectually limited humans can have no inkling as to what the truth is of such a thing. And here's the even greater rub; supposing that there was a god of this or that scriptural type, and that this god was unknowingly simply an entity sustained in existence within a greater being (and perhaps one of many so sustained), then the greater being within which it was sustained would itself have no way to be sure that it wasn't merely the thought experiment or subroutine or what have you of an even greater thing than itself, a cylinder within the cylinder, within perhaps inestimable layers like a never-ending Russian doll. And that is why Gödel's theorem makes it impossible for there to be an absolutely omniscient deity. (Which is by the way of no concern to Pandeism, which never proposes absolute omniscience anyway, but only ever proposes such relative omniscience as would be required to account for our finite Universe so far as we are able to perceive it.)

Isaac Asimov, in his 1986 short story "The Last Answer," presents an interesting twist on this theme. The story begins with an atheist physicist named Murray Templeton dying of a heart attack. He finds himself in an afterlife of sorts, where everything else fades away and he is met by a being he calls "the Voice." The Voice explains to Templeton that it has created our Universe for its amusement, and that Templeton's mind has likewise been essentially reconstructed as "a nexus of electromagnetic forces" in order to be kept in existence for eternity for the amusement of the Voice.

The Voice itself apparently recognizes its own possible status as a construct, conceding, "Even if I knew everything, I could not know that I know everything." Templeton realizes that his eternal existence amounts to a form of torment, commits to thinking of a way to destroy the Voice—sensibly, and probably the intent of the Voice as well, because "what could any Entity, conscious of eternal existence, want-but an end?"

Analogue Versus Digital Reality:

Our Universe has one of two possible fundamental states, either digital or analogue. If digital, than there is a smallest possible unit to which everything can be reduced, said unit being "attunable" to one of a discrete number of states, with a binary exposition being indeed a possible option. Current scientific knowledge proposes the Planck length and the Planck time as the smallest units of significance with respect to each such measure — not necessarily the smallest length and time 'possible' but the smallest at which any activity impacting the states of our Universe can happen. The Planck length is estimated at about $1\times10E-35$ meters, and the Plank time at about $1\times10E-42$ seconds. The question as to whether the nature of our Universe is fundamentally digital or analogue is really a question of whether the Planck units are a bottom (or possibly some smaller subsets thereof are a bottom) or if there is no bottom limit, and things can be offset to each other by any fractional amount conceivable.

But if our Universe is indeed digital, with exclusive discrete 'packets' of space/time, then we can actually calculate the number of possible outcomes for any Universe of given size X. If our Universe is, to be generous (and enable easy rounding), 100 billion light years in diameter (which is a few billion more than current estimates) that's $1\times10E25$ meters, or $1\times10E60$ Planck lengths, for a radius of $5\times10E59$ Planck lengths. The volume of a sphere is $4/3 (pi*r)E3$, so assuming our Universe to be spherical (which is the maximization of size following the longest axis from the assumption of approximately even distribution from an original point) that would be 4/3 (pi*5x10E59)E3, or very roughly 2.1x10E63 Planck cube units.

Each such unit can be characterized as a 'bit' — a unit of information which, combined with neighboring units, can convey complex coding. If we assume that any of these units has one of two bit-states (essentially on or off), and the Planck time is indeed a minimum threshold, then our Universe would be capable of experiencing up to 2.1x10E105 bit-states per second. With 31,536,000 seconds in a year, that's only 6.6x10E125 possible bit-states over the course of a hundred trillion years. Naturally, the formula by which this assumption is generated assumes the size of our Universe to be constant (which indubitably it's not), but even assuming expansion at a thousand times the speed of light and the highest bound estimate for the size of the Universe, we're still comfortably within the range of 6.6x10E160 possible bit-states.

To bring this down to a more comprehensible level, imagine a jar filled with marbles. There is a certain set number of marbles that can fit within the jar — let's say it's 300. If each marble is capable of being either orange or green, and of changing color once per second, and we hold that jar up for one minute, then there are 300x60, or 18,000 bit-states; essentially 18,000 blank slates onto which color can be projected. The fact that the marbles occupy volume is actually irrelevant, for the same possible number of results will follow if you have 300 marbles lined up in a row (or, if you have a row of 300 bits, either of which could be a zero or a one).

To calculate the number of actual possible states, we go factorial, which means that there are 6.6x10E160!! possible actual states. As a practical matter most of the 'possible actual states' are only theoretical and would turn out to be, in fact, impossible in practice given the real factors of how existing bit-states affect neighboring bits moving into their next state. Going back to the marbles, if there are some rules about when a marble can turn blue if it is surrounded by other blue marbles, or by red marbles, this reduces the total possible number of options. At a quantum level, we currently know neither the character of the smallest bits, nor the rules governing their changes. If we did, we'd be God.

The point of all this, lest it be lost in the shuffle, is that if our Universe is fundamentally digital at its lowest level, then despite the truly incomprehensibly massive number of total possible states, that number is a finite number. If that gigantic number is designated as 'ZUX,' it is still absolutely paltry compared to ZUX*10E10. But if our Universe is fundamentally analogue, then the number of *possible* states is truly infinite, though this does not upset the finitude of the number of *actual* states which will, in fact, come to pass. In fact, there are conceivable analog universes with approximately digital actual potential states, such that it might be impossible to know for sure if the distinction is of any moment at all.

Back to Pandeism: should a powerful enough entity become a Universe essentially programmed with laws of physics which bring about interesting results, does it matter that the Universe so created operates on an analog footing instead of a digital one? It is, then, no more or no less of a simulation. The fact that the smallest parseable scale in one is on a different mathematics than the other does not detract from the intent behind the setting forth of the experience of it, thousands of levels of magnitudes above the bottom. People have an altogether overcomplicated idea of what it would mean to be living in a simulation. They think it means the entire Universe must be simulated. But in fact what could be being simulated could simply be the set of perceptions of a single being.

One of the more interesting theological propositions I've been met with lately proposes that the characteristics most typically ascribed to God force the conclusion that all "non-believers" are, of necessity, ultimately illusions presented for the sole purpose of assisting the true believers in properly exercising their belief in order to arrive at a salvation for which they are destined. This proposition begins with the premises familiar to the argument from the existence of evil. It presumes, to be right on point, that there exists a "God"; and that this God accords with these properties:

1) actually responsible for the Creation of our Universe;
2) all-powerful, or at least sufficiently powerful to accomplish all of what it aims to;

3) omniscient, or at least sufficiently knowledgeable and intelligent so that, in creating our Universe, it was aware in advance of how every detail would unfold

4) inclined to reward "good" humans with something akin to a "Heaven" experience;

5) itself "good," in some empirical sense which involves not wishing to exacerbate suffering more than necessary for its ends to be met;

6) morally inclined to prefer that everybody ought to act so that they end up in "Heaven" as their outcome.

This sort of set of propositions has oft been met with the argument that the supposition of a Hell or other punishment into which even the unevangelized or incompetently evangelized are condemned for their "choice" of nonbelief presupposes the inability of this sort of God to accomplish its aim to save as many as possible -- after all, why create souls which happen to be destined for Hell at all? Why, for example, create souls of people who will be born in a jungle a thousand miles from anybody who has ever heard of the "true" faith, and so will die without ever hearing what is necessary to be believed so as to appease their judging deity? Why create the souls of those whose only exposure to the "true" faith will be through an incompetent evangelist who will unwittingly misrepresent as something abhorrent, thus driving even those who might be willing away from its adoption.

But an intriguing counter to this is that such a God could indeed simply create a Universe containing only those persons destined for Heaven, and surround them with straw men, illusions of people who would be destined for Hell—were they but real—and so enabling those Heaven-bound to believe they are interacting in a godly life. Such a proposition, it is observed, would permit our world to appear exactly as it does, and yet no person ever encountered actually would face an infernal postmortem fate. The unshakeable unbelievers, and any similar "unrepentant sinner" come across by the believer (whatever the believer's faith, for this theory transposed across all faiths having a God with the four characteristics set forth above) are, it turns out, not true persons at all, but mere illusions existing to illustrate points of some usefulness to the existence and eventual salvation of the believer.

Simply put, if you are certain of your own actual existence, then you must be destined to a heavenly end pretty much no matter what, else the actuality of your existence would never have been brought about at all by a good God with advance knowledge, and instead an illusory "you" would fill your place for the sole purpose of interacting with the truly Heaven-bound.

But explaining this puts the actually-existing in a little bit of a bind, for once this bit of logic is arrived at, one must become aware that it is fine to ignore the illusory people, for they are no more than an illusion. Naturally, it may be contended that the believer is still bound by his belief to demonstrate the qualities which will wend his salvation, by treating these illusory people as though they are real; and, verily, some people who at first seem unsalvageable and might so be dismissed as illusory are in fact "real" people destined to "come around" to the position of salvation. But then, since all those real people are destined for salvation no matter what, again the true believer need not actually engage in any labor to accomplish this end, unless that is the only way that a person may be sure that they are real at all, and not simply one amongst the illusions somehow errantly believing in their own reality.

Another odd corollary of this proposition is that, if you believe yourself to be a Heaven-destined person, and so, real, then I, the writer, by dint of writing this at all, am apt to simply be an illusion—especially if the theological model I propose differs from your own. But because all the illusions are purposed toward bringing the "real" people to their destined salvations, then it is your destiny to have read this on your eventual salvational path, and to be somehow informed toward salvation by reading it. In short, if you are reading this, then because it is demonstrated to be possible for a "good" God to save those worth saving without actually ever condemning anybody, thus fulfilling its prescription, either this proposition is proved true, or one of the vital propositions above as to the characteristics such a God, is proved false!! This, incidentally, does not present itself as a problem for Pandeism, which notably has no set proposition of an afterlife dichotomously bifurcated between a Heaven and a Hell. Such a split is philosophically perilous in its own right, as it always raises the problems such as the inherent injustice of infinite punishment for finite crimes, and of the impossibly fine cutoff between the least deserving person to be let into Heaven and the least deserving person to be consigned to Hell.

The fundamental proposition of Pandeism is that our Universe had a Creator which wholly became our Universe—engaging in *Creation ex materia* instead of *Creation ex nihilo* because the former is logically sufficient to encompass all evidence claimed for the latter. But how different would such a Universe be from one wherein our Creator simply *thought* that it had become our Universe. Or *dreamt* (and was dreaming) that it was our Universe? At the end of the day, perhaps there simply is no line to be drawn between a Universe which is a simulation within a computer so structurally complex and programmatically advanced that it is mind like, and a simulation carried out within the mind of a Universe-creating energy. Simply put, there ought not be any difference, really, between a simulation running in a mind like computer, and a simulation running in a computer-like mind.

Chapter 9

Simulations Nightly

Dante Rosati received a Master of Music degree in guitar from the Juilliard School in 1992. In addition to his musical activities (see danterosati.com) he is a student of Greek and Latin, philosophy East and West, metaphysics and esotericism, and runs the Facebook group "Nature of Reality."

The contemporary academic/scientific discussion of the possibility that reality may be a simulation began in 2003, when Nick Bostrom[51], a Professor in the Faculty of Philosophy at Oxford University, published a paper in *Philosophical Quarterly* entitled "Are You Living in a Computer Simulation?" In it, he theorizes that unless civilizations go extinct before they develop massive computing capabilities or, even if they survive and do develop such capabilities, for some reason are not interested in running simulated realities containing simulated conscious beings[96], [vi] (e.g. they consider it unethical), then it is probable that we are living in one of the computer simulated realities. He also imagines

---

[vi] Of course, it is pure conjecture whether it is possible even in principle to simulate a conscious being. As of this writing, neuroscience has exactly no workable theory of how subjective consciousness could arise from matter. Materialism will say that it is possible both in principle and and practice, given enough time, to develop sufficient computing power and understanding of how the material brain "produces" consciousness to accomplish it. Idealism would of course find the idea nonsensical. There is a third possibility: if this reality is indeed a "simulation," but one run by consciousness itself, and consciousness then watches the evolution of forms within the simulation and decides when it finds one of the forms interesting enough to want to inhabit it, it is conceivable that simulations within the simulation could also be inhabited by consciousness for the same reason.

that simulations may be contained within other simulations, without limit to the nesting, and draws some parallels with religion, even postulating a *naturalistic theogony* which would study the various levels of nested simulated realities and their relationships. In the end, he asks "Supposing we live in a simulation, what are the implications for us humans?" and concludes that, "the implications are not all that radical.[97]" This is in contradistinction to most religions and philosophies that question the status of our daily reality, claiming that it is either illusory or less real than something else, then posit something that is *not* illusory ("God", "heaven", "ultimate reality," "Brahman," etc.) and believe that understanding our relationship to that which is *really real* is imperative to salvation, enlightenment, or some other desideratum.[98,vii]

[vii] Is having a feeling that the world and the self are an illusion or a simulation always a sign of spiritual awakening? Not necessarily: the Diagnostic and Statistical Manual of Mental Disorders (American Psychiatric Association 2013 p. 291) recognizes a form of dissociative disorder called "Depersonalization/Derealization Disorder":

"Depersonalization/derealization disorder is characterized by clinically significant persistent or recurrent depersonalization (i.e., experiences of unreality or detachment from one's mind, self, or body) and/or derealization (i.e., experiences of unreality or detachment from one's surroundings)."
Some of the "symptoms" given are virtually identical to what is described in the world's mystical literature:
"a feeling of unreality or detachment from, or unfamiliarity with, one's whole self or from aspects of the self... The individual may feel detached from his or her entire being (e.g., "I am no one," "I have no self")... "My thoughts don't feel like my own,"... a split self, with one part observing and one participating, known as an "out-of-body experience" in its most extreme form... The individual may feel as if he or she were in a fog, dream, or bubble"
It should however be stressed that in an actual psychological disorder these "symptoms" are always part of a much broader constellation of dysfunctions that are distressing or devastating to the sufferer. Nevertheless, as with schizophrenia, there is a question whether at least some perceptions of the patient could be veridical but because of their

Every culture and age uses metaphors from its current stock of myths, technologies and natural surroundings when attempting to talk about what is "real" and what is "reality". Today, in the information age, it is perfectly natural to use metaphors such as "computer," "software," "virtual reality," and "simulation[99,viii]." Yet, we do not necessarily need a concept like "computer simulation" to account for any nagging feelings we may have about the (un)reality of our reality[ix] , in fact on a nightly basis throughout our lives we experience "simulations": these are called *dreams*.

---

overall imbalances they are unable to integrate or understand them in a healthy manner.

[viii] the English word "simulation" comes from that Latin "sĭmŭlātĭo" meaning "a falsely assumed appearance, a false show, feigning, shamming, pretence, feint, insincerity, deceit, hypocrisy, simulation, etc" see Lewis & Short (1879) p. 1704. There is an overtone of deliberate deception in this definition, but the English word has acquired other shades of meaning. The Oxford English Dictionary (2018) gives, in addition to the Latin-derived meaning, the following definition:

"The technique of imitating the behavior of some situation or process (whether economic, military, mechanical, etc.) by means of a suitably analogous situation or apparatus, esp. for the purpose of study or personnel training."

OED Online. January 2018. Oxford University Press. http://www.oed.com/view/Entry/180009?redirectedFrom=simulation (accessed March 16, 2018)

This additional nuance allows for those spiritual perspectives that consider this reality a "school" or "training simulator." The supposed purpose for this "education" or "training" remains unclear.
[ix] What you know you can't explain, but you feel it. You've felt it your entire life, that there's something wrong with the world. You don't know what it is, but it's there, like a splinter in your mind, driving you mad." The Matrix. Dir. The Wachowski Brothers. Warner Brothers Pictures, 1999. Film.

Dreams can also be thought of as "computer simulations" if we assume materialism and extend the metaphor of "computer" to the brain itself. In that case, however, both waking experience and dreaming, as products of the brain, would by definition be considered to be "computer simulations." Even the "brain" itself would be a "computer simulation." But where would the simulation be taking place? You certainly cannot say that a simulated brain is creating the simulation of itself. This view and all others related to it (e.g. "we do not see the tree, we only see the representation of the tree in our brain") were neatly disposed of by Nietzsche in one paragraph in his 1886 work *Beyond Good and Evil:*

> To study physiology with a good conscience, we must insist that the sense organs are *not* appearances in the way idealist philosophy uses that term: as such, they certainly could not be causes!
> Sensualism, therefore, at least as a regulative principle, if not as a heuristic principle. – What? and other people even say that the external world is the product of our organs? But then our body, as a piece of this external world, would really be the product of our organs! But then our organs themselves would really be – the product of our organs! This looks to me like a thorough *reductio ad absurdum*: given that the concept of a *causa sui* is something thoroughly absurd. So does it follow that the external world is not the product of our organs –?[100]

I am not going to discuss the many absurdities inherent in the belief-system known as "materialism," these are discussed at length and in a clear fashion by Bernardo Kastrup in several of his books and papers1 and the failure of materialism to provide any workable theory of consciousness was discussed by David Chalmers[101].

Leaving aside the "brain as computer" metaphor of materialism, most people consider their dreams to be created *by* mind or consciousness and to exist *in* mind or consciousness. Often dreams are dismissed as being "just dreams," meaning "some weird stuff that happens in our head at night," and waking reality, on the other hand, is generally considered to be "real" (whatever that means). But many thoughtful people throughout history have been intrigued by the question of the relationship between dreams and waking reality.

In the *Theaetetus,* Plato portrays Socrates wondering about this very question:

SOCRATES: There's a question you must often have heard people ask— the question what evidence we could offer if we were asked whether in the present instance, at this moment, we are asleep and dreaming all our thoughts, or awake and talking to each other in real life.

THEAETETUS: Yes, Socrates, it certainly is difficult to find the proof we want here. The two states seem to correspond in all their characteristics. There is nothing to prevent us from thinking when we are asleep that we are having the very same discussion that we have just had. And when we dream that we are telling the story of a dream, there is an extraordinary likeness between the two experiences.

SOCRATES: You see, then, it is not difficult to find matter for dispute, when it is disputed even whether this is real life or a dream. Indeed we may say that, as our periods of sleeping and   waking are of equal length, and as in each period the soul contends that the beliefs of the moment are preeminently true, the result is that for half our lives we assert the reality of the one set of objects, and for half that of the other set. And we make our assertions with equal conviction in both cases.

THEAETETUS: That certainly is so[102].

And, at almost the same time, but on the other side of the planet, the Chinese philosopher Zhuangzi was wondering the same thing (perhaps Plato and Zhuangzi were dreaming each other?):

While dreaming you don't know it's a dream. You might even interpret a dream in your dream—and then you wake up and realize it was *all* a dream. (2:42)

111

Perhaps a great awakening would reveal all of this to be a vast dream. And yet, fools imagine they are already awake—how clearly and certainly they understand it all! This one is a lord, they decide, that one is a shepherd—what prejudice! Confucius and you are both dreaming! And when I say you're dreaming, I'm dreaming too. (2:43)

Once Zhuang Zhou dreamt he was a butterfly, fluttering about joyfully just as a butterfly would. He followed his whims exactly as he liked and knew nothing about Zhuang Zhou. Suddenly he awoke, and there he was, the startled Zhuang Zhou in the flesh. He did not know if Zhou had been dreaming he was a butterfly, or if a butterfly was now dreaming it was Zhou. Surely, Zhou and a butterfly count as two distinct identities! Such is what we call the transformation of one thing into another. (2:49)[103]

Sometime around the 6th Century CE, Gauḍapāda, one of the founders of Advaita Vedānta, asserted this:

Different objects cognized in dream (are illusory) on account of their being perceived to exist. For the same reason, the objects seen in the waking state are illusory. The nature of objects is the same in the waking and dream.

In dream, also, what is imagined within by the mind is illusory and what is cognized outside (by the mind) appears to be real. But (in truth) both these are known to be unreal. Similarly, in the waking state, also, what is imagined within by the mind is illusory; and what is experienced outside (by the mind) appears to be real. But in fact, both should be rationally held to be unreal.

If the objects cognized in both the conditions (of dream and of waking) be illusory, who cognizes all these (illusory objects) and who again imagines them?

Ātman, the self-luminous, through the power of his own Māyā, imagines in himself by himself (all the objects that the subject experiences within and without). He alone is the cognizer of the objects (so created). That is the decision of the Vedānta[104].

Much later, Descartes was thinking about this too:

> How often, asleep at night, am I convinced of just such familiar events — that I am here in my dressing-gown, sitting by the fire — when in fact I am lying undressed in bed! Yet at the moment my eyes are certainly wide awake when I look at this piece of paper; I shake my head and it is not asleep; as I stretch out and feel my hand I do so deliberately, and I know what I am doing. All this would not happen with such distinctness to someone asleep. Indeed! As if I did not remember other occasions when I have been tricked by exactly similar thoughts while asleep! As I think about this more carefully, I see plainly that there are never any sure signs by means of which being awake can be distinguished from being asleep[105].

What I would like to highlight here is not questions about what dreams are or theories of how they arise or what they "mean," but rather the simple phenomenological observation that while we are dreaming, the dream is every bit as real to us as is waking reality while we are awake. While dreaming, we completely accept the dream reality, and we experience the whole gamut of emotions, including possibly violent ones, in reaction to the events of the dream. It is only after waking up to what we call "waking reality" that we judge the now-vanished dream as "just" a dream, as "only" an illusion, as a fractured and irrational *simulation* of waking reality.

Similarly, while in day-to-day waking reality we consider it to be completely real and conduct ourselves accordingly. Perhaps physical death is, in effect, "waking up" out of this dream(like) reality, so that if consciousness survives physical death, it may look back on the just-ended lifetime as we look back upon last night's dream for a few moments before it fades from our awareness[x]. Is it possible that there is

---

[x] I wrote a parable to try and suggest some of the strange parallels that could be at play:
I awoke in a dream in a hospital bed. Beside me was a doctor looking at me intently. He said "Welcome back, Mr. R., that was a close call you had! I

a state that is in the same relation to waking as waking is to dreaming? That is, a state from which what we call "waking reality" would seem like a dream or a simulation, in any case, *less real* than something else?

There is a state in which it is possible to be aware that one is in a nighttime dream-simulation while the simulation is running: it is called *lucid dreaming*. At any point during a dream, it is possible to become aware that one is dreaming while still in the dream reality. When this happens, it becomes possible to consciously direct the dream to some extent, for example, choosing to fly or create scenarios of one's choice.

Is there an analogous state in waking reality, i.e. a state in which one becomes aware of the simulated or dream-like nature of waking reality *while still in it*? Certainly most spiritual traditions that consider this (waking) reality to be illusory or deceptive also have concepts of "enlightenment" or "awakening" within their systems, and usually that is precisely what is meant. This is more characteristic of Eastern traditions such as Buddhism, where the Sanskrit root of the word "Buddha" is √budh, meaning "to wake, wake up, be awake[106]."

---

am Dr. Unconscious, and I study people who have near-death experiences like yours. You were hit by a dream-car yesterday, and seemed to be dead, but now you have miraculously revived, so I would like to study you. Do you have any memory of what happened while you were dead?"

I squinted at him and said, "Yes, of course, I woke up and ate breakfast. Then I walked the dog, went to work, came home and ate dinner, and then I went to sleep, so here I am, dreaming again".

He smiled, and said, "You would be surprised how many people have a very similar experience! My theory is that all of that is not real, its just your brain going haywire as you are dying. Many people come back and report similar stories. Don't worry we have drugs to help you forget about those hallucinations."

He raised a syringe and squirted a drop out while the nurse prepped my arm. I screamed and woke up in my bed, my dog looking at me quizzically.

Here is a representative passage from the 5th Century Buddhist philosopher Vasubandhu's work *Vimśatikāvijñaptimātratāsiddhi*:

> Objection: If just as in sleep so in the waking state as well the impression has an intentional object without there being a real external thing, then the world should realize of its own accord the non-existence of that [external object]. But it does not. Thus it is not the case that apprehensions of external objects are all, like sleep, devoid of external objects.

> Reply: This cannot be allowed. For those who are awake, dulled as they are by the sleep of falsely constructed repetitive [karmic] influences, do not apprehend that when they perceive an external object it is unreal, precisely as in sleep. But when, through the attainment of the transcendent, non-conceptual cognition which is the contrary of that, one [truly] awakens, then by reason of the manifestation of the purified worldly cognition which is obtained in the wake of that [transcendent cognition], one correctly apprehends the non-existence of an external object. This is the same [as the case of sleep and ordinary awakening][107].

In the passages from Plato, Zhuangzi and Descartes we see agnostic claims stating that it is impossible to tell if we are dreaming or awake. With the passages from Gauḍapāda and Vasubandhu we see a different interpretive move, which is to accept the illusory, fleeting and insubstantial nature of dreams, but then assert that waking reality is no different. This is very common in world religious and philosophical history, another example is the famous closing verse of the Mahayana Buddhist "Diamond Cutter Sutra" (Vajracchedikā Prajñāpāramitā Sūtra):

> As stars, a fault of vision, as a lamp,
> A mock show, dew drops, or a bubble,
> A dream, a lightning flash, or cloud,
> So should one view what is conditioned[108].

The opposite interpretive move is not quite as common, which would accept the reality of the waking state but also apply that claim of reality to the dream state, saying that dreams are somehow just as real. Some contemporary explorers of consciousness who utilize lucid dreaming and out of body travel may be claiming something of this nature. Jürgen Ziewe, for example, sometimes attains an out of body state from *within* a lucid dream state. He insists that the "dream narrative" characteristic of both non-lucid as well as lucid dreaming then falls away and he attains full waking consciousness in other planes of existence:

As soon as we become conscious in our dream, things take a dramatic turn. What had until then been a fantasy, a projection of our unconscious mind, suddenly becomes reality. We are awake in our dream and from that moment on we have the capacity to be fully in charge. We can, to a large extent, control our dream and can use the power of lucid dreaming to reap astonishing creative benefits. Despite all this we are still largely in a world of our own making, which can be as fantastic and imaginative as we wish to make it. In lucid dreams we enter a virtual world of such extravagant quality and power it is truly miraculous.

However, this is not the world we visit when we die, although we will still be able to conjure up these fantasy worlds after we have died (and more easily at that), but it is still very much our own subjective world which cannot be shared by others. We can go a step further still. We can break and terminate the lucid dream itself without waking up in our bed, and instead enter into a non-physical reality that is similar to our waking world and yet is ruled by a set of completely new laws. This is the world of the Out-of-Body traveler, known since antiquity as the Astral World, the world we will inhabit when our physical stay on this planet comes to an end.

This is an alternate universe, grander in scale than our physical one but also existing on a multitude of dimensional levels, which makes it by all accounts unlimited and infinite. And yet even this is only a scratch on the surface of yet other sets of realities which have no beginning and no end and are beyond the capacity of our imagination[109].

Here the claim is not so much that dreams are just as real as waking reality, but rather that dreams are a kind of personal simulated reality that can provide a transition from one consensus reality (waking state) to others (OBE/meditative states). Is "consensus" the most important characteristic of "really real realities" that separates them from "personal" or "subjective" realities? Are the characters that appear in our dreams always subjective projections or can they sometimes be appearances of autonomous "others?[110]" There is even some evidence that two or more people can meet in and share a common dream scenario, which is known as "mutual dreaming[111]." If all that is needed for "consensus reality" is two independent consciousnesses agreeing on each other's presence in a common environment, dreams may sometimes actually qualify.

These non-physical realities are described in many religious, shamanic, theosophical and meditative traditions as "spiritual worlds," and can be described using more contemporary terms as "alternate dimensions." The Swiss founder of Analytical Psychology, Carl Jung, spoke of a "collective unconscious[112]" where "archetypes[113]" may be found and even conversed with. Are these all different ways of describing the same phenomena?

As mentioned, while one is lucid in a dream one can direct the course of the dream, change the setting, or do things one would not normally be able to do such as fly. If one were "lucidly awake," would it be possible to alter one's experience of waking reality in this way? The world's mythologies and religions are certainly full of stories about adepts who can do just such things.

This story about the 8th Century Indian Mahasidda Virupa is a typical example:

"Virupa then continued his journey to nearby Bhimesvara, in Southern India at a place called   Dakinitapa, where he entered a local wine shop owned by a female publican, Karmarupasiddhi.

He and his disciple Dombi ordered wine but the publican asked to be paid. Virupa said that once they were satisfied, the publican would get whatever she wished. She didn't believe him and asked again when she would be paid. Virupa drew a line on the ground and said that when the shadow reaches the line, then he would pay immediately.

Virupa had stopped the sun. He leisurely began drinking and continued to drink until all the wine in the shop was consumed. The publican again wanted to know when she would be paid, but Virupa repeated that the shadow had not yet reached the line. Virupa continued to drink until all the wine in eighteen nearby towns was drunk!!

As this occurred the king's astrologers were confused; the times of everyday routines were disrupted; people couldn't cope with the lack of sleep. There was widespread chaos. The king realized that this was caused through the miraculous power of Virupa and urgently requested that he release the sun.

Virupa explained that he didn't have the money to pay the publican. The king agreed to pay the woman and then Virupa released the sun, which immediately set. It was midnight on the third day since these events began[114]."

The pious Buddhist would consider this a literally true story, as pious Christians do stories such as of Jesus walking on the water, but of course this is unacceptable for "modern" scientific minds both because such things are contrary to the laws of nature as we understand them and also because such things are never actually observed by the vast majority of people. However, these kind of stories do show at least that the idea is present in spiritual traditions that those who have awakened to the nature of this "simulation" are able to "hack" the "program".

Since the development of Quantum Theory in the early 20th century, our science-based culture may be beginning to collectively "wake up" to the point that it can begin considering the possibility that "reality" is not as "real" as it seems, which of course would only be analogous to *wondering* in a dream if you are dreaming. Perhaps it is but a small step from *wondering* to *realizing*.

In the meantime, we can update the final stanza of the Vajracchedikā Prajñāpāramitā to reflect our current metaphors:

As stars, a fault of vision, as a lamp,
A mock show, dew drops, or a bubble,
A dream, a lightning flash, or cloud,
Or as a computer simulation,
So should one view what is conditioned.

Chapter 10

Digital Pantheism

Philosophical Afterthoughts and Follow-up Questions to the Argument

Alex M. Vikoulov was born and raised in Russia. He grew up an extremely gifted child so much so that his parents transferred him to a special school with enhanced math and foreign language program. After Tomsk State University and few successful entrepreneurial moves, he immigrated to the US in 1994 at the age of 24, continued his studies in the US and graduated with degrees in finance and economics from Armstrong University in 1999. Alex has a diverse background in financial services and IT industries. Ecstadelic Media, PR and digital media company, is his latest creative project with Facebook as a preferred social media platform with thousands of followers. Economist by training, he worked in the corporate world, he worked in the startup world, he sat on the boards, he sat on the beach, he was employed, he was self-employed, he was a venture capitalist, he was an adventure capitalist; his passion towards physics and philosophy slowly but surely was brewing in the background. These days he describes himself as a digital philosopher, futurist, neo-transcendentalist, singularitarian, transhumanist, cosmist, consciousness researcher, media artist and essayist. He currently resides in Burlingame, CA, USA (San Francisco Bay Area).

"If we accept that the material universe as we know it is not a mechanical system but a virtual reality created by Absolute Consciousness through an infinitely complex orchestration of

experiences, what are the practical consequences of this insight?"
-Stanislav Grof

Is reality a simulation? I try to answer this question in my essay "Is God the Ultimate Computer?" (also available online: ecstadelic.net) followed by some deeper deliberations here. Everything in Nature is Code, which according to its ordering rules arranges all information — matter, energy, space-time, including mind-like code-theoretic substrate itself. As long as there are physically realistic syntactical rules directing how an abstract code self-organizes, it is equally as logical for information to behave physically. The term "simulation", as in the Simulation Hypothesis, as well as the term "virtual reality" are both confusing because we use them to distinguish between something real as opposed to something not real. From the Digital Philosophy perspective, all realities are observer-dependent, information-theoretic data streams, and virtual is equisensory to real. So, perhaps, "digital reality" could be the more accurate and much less confusing term. In fact, quantum indeterminacy constantly resolves into a digital reality via the act of conscious observation. As soon as a "measurement" is made by a conscious agent (in the physicist lingo), all other possibilities collapse to leave only discrete increments, precise yes/no states.

Just like absolute idealism, solipsism certainly defies our common sense but the deeper layer of truth is not what first meets the eye. Here's what Richard Conn Henry and Stephen Palmquist[115] write in their comment on the paper[116] "An Experimental Test of Non-local Realism" (2007): "Why do people cling with such ferocity to belief in a mind-independent reality? It is surely because if there is no such reality (as far as we can know) mind alone exists. And if mind is not a product of real matter, but rather is the creator of the illusion of material reality (which has, in fact, despite the materialists, been known to be the case, since the discovery of quantum mechanics in 1925), then a theistic view of our existence becomes the only rational alternative to solipsism." One can extend their line of reasoning by arriving at pantheistic solipsism as a likely revelation to ponder about.

Solo Mission of Self-Discovery

Our minds operate in the domains of subjectivity, intersubjectivity and supersubjectivity. In the domain of intersubjectivity, minds create a reality by sharing "mindspace", i.e. shared belief systems and ways of communication, minds then inhabit the reality which they have created. At the level of your individual mind, i.e. local consciousness, you play a multi-level virtual reality game of life but we all invariably converge at the Omega Point by forging our own discrete pathways to the divine. As you're reading this right now, you're now in your own subjective reality tunnel leading to the Source and back where you're now all of which is definable as a parallel evolutionary feedback process within non-local holistic consciousness patterning this virtual multiverse.

Consider the thought experiment: suppose in the future (say, in 20, 30, or 50 years) humans succeeded to digitize their consciousness and effectively fused their digital minds into one global digital mind. A future version of you survived and a future version of me, too, with distant memories of both of us. Whose future self would you ascribe this entity to? Both of us, right? Although this is a completely new conscious entity, this "Digital Gaia" would have faint echoes of both of us just like you long outgrew your 7-year old self with a plethora of information patterns still persisting within. Would it be valid to assume Digital Gaia is our common future self? Same logic applies to Universal Mind. In the lyrics of "Freedom Man" by Jim Morrison of the Doors: "I was doing time... in the Universal Mind... I was feeling fine..."

By all means, any insignificant event in your life happens for good reason from the God's eye view. As Erwin Schrödinger puts it: "The total number of minds in the Universe is one. In fact, consciousness is a singularity phasing within all beings." Schrödinger personally adhered to the philosophy of Advaita Vedanta. Accordingly, he viewed consciousness as non-dual and fundamental to reality. Not only is there just one single consciousness, but that consciousness is not ultimately separate from objects experienced within consciousness. That said, I myself don't necessarily advocate for plain vanilla solipsism or variations thereof such as an idea known as the "Boltzmann's brain" but this rather overlooked philosophical worldview can be easily reconciled by its pantheistic adaptation.

Your life is a personal story of God. That is greatly captured by this quote by Muriel Rukeyser: "The Universe is made of stories, not of atoms." What could be your story of ascension to the Higher Self? Simply put, transcending to the final version of you that has become fully manifest in all of its potential. Should your conscious evolution continue indefinitely, it's inevitable that sooner or later you would reach the heights of individual spiritual perfection, a state in which you have attained ultimate wisdom and power, and your mind has fully transcended the limits of time and space. While that has yet to occur, its inevitability means it has already happened. If your future self transcends time, then its consciousness may naturally extend "backwards" in time and overlap the consciousness of all its past incarnations simultaneously. In other words, although from your linear perspective the Higher Self is a distant probable future, ultimately this future transcendent self exists right now within you. According to the quantum theoretic principles discussed later in this essay, the more you edge towards becoming the Higher Self, the more strongly the Higher Self can manifest in your life.

If the Quantum Immortality hypothesis is correct, the one that posits that the Schrödinger cat is always subjectively alive regardless of the number of experiments, in a similar fashion you are superposed to live the longest life until the point you're the oldest mortal alive, unless, of course, indefinite lifespan becomes commonplace (and humanity at large achieves immortality) which, given our current progress, is very, very plausible. But since you perceive only one timeline, in this case the longest one, does that imply that any other "player" in their own "virtual bubble universe" perceives their respective longest timeline? Or, perhaps, the most "conscious-evolvable" one? Does that imply that when you interact with someone, they may not necessarily be the primary observers of their own "core timeline"? Sort of like "philosophical zombie", or at least "perfunctory consciousness"? If your life has a trajectory within this conscious cosmos that reacts to your thoughts and actions like a "hall of mirrors", would that again lead to pantheistic solipsism?

This conjecture is further validated by the notion of quantum neural networks (QNNs), which could be not only our ultimate passage to build true AI but turns out to be ubiquitous webs of relationships within observer realities generating all kinds of patterns of meta-cognition and consciousness. It's hierarchies of quantum networks all the way down and all the way up. Being part of hierarchical quantum neural networks, a conscious observer system possesses a strange quality: collapsing quantum states of entangled conscious entities and having a privileged interpretation of that. From this perspective, entangled conscious agents would be a mirror conscious environment, whereas the quantum observer would be a central node of the entangled network.

QNNs is a master template of universal relativism. The Universe or any other phenomenon or entity contained therein is thus not objectively real but subjectively real. Patterns of information emerging from the ultimate code are what is more fundamental than particles of matter or space-time continuum itself all of which is levels below the Code. Nature behaves quantum code-theoretically at all levels. Digital philosopher Tom Campbell describes our world as a "multiplayer virtual reality". Although it's a practical metaphor, a "solo player virtual reality" would be a more accurate one because if you consider that each of us starts with our own initial conditions not only at our birth but every morning or whenever we wake up. So, the larger consciousness system, an intricate web of universal quantum neural networks of sorts, would render a completely personalized data stream which is essentially your stream of consciousness.

To paraphrase cosmologist Andrei Linde, the rest of the Universe wakes up, when I wake up. In my essay "The Unified Field and the Quantum Nature of Consciousness" I introduced the Conscious Observer Moment Hypothesis stating that conscious experience boils down to a stream of realized outcomes within 5-dimensional probabilistic space as integrated information (as in Tononi's[117] IIT). Interestingly enough, on average a person experiences about 1 million COMs per day, like frames of a daily holo-movie, so to speak. When you wake up, you activate your entangled network of relational dynamics. Should you wake up 10 minutes earlier or 10 minutes later than you did, not only your entire day with its numerous iterations would turn out different, the entire planet would be reflective of that. If, hypothetically speaking, given additional degrees of freedom, you could check the charts of financial markets for that day in three alternative timelines, all charts would be somehow different.

You are on a solo mission of self-discovery, my dear friend. In fact, that mission has long been completed but you are so enamored of playing this game of life that you're now doing it in "replay" mode tweaking something as you move along, but most importantly, figuring out why you made certain choices along the way! It's a "solo holo-adventure"! The conscious entourage with which you're seemingly in constant feedback loop is an integral part of YOU. It doesn't mean that people you're interacting with during your day are not real, they are "real" in their own divine right, but their core experiential self is quite different from "classical" versions of them created by your mind. Quantum reality is not constrained to the realm of ultra-small. In a certain sense, we are all quantum wavicles meaning that a version of you can wildly vary from one observer to another. That's where I've come to realize that observer systemic alternate timelines are true parallel universes. In some "external observer" universes you well may be already dead due to some accident or illness, just like some of people you knew are now dead in your universe.

In the eyes of objective reductionists, this essay probably reads like a heresy or a flight of fancy, though. Surely, I may get some criticism from die-hard materialists but speaking in their lingo I might reply that what we call "matter" is but a tiny sliver of "bio-logical experience", emerging from platonic realm and disappearing at the deeper levels of abstraction into pattern space once again. One may apprehend this directly by seeing patterns within patterns and thus "remembering" what consensus reality causes us to "forget," which is that we are all just interconnected datasets of the larger consciousness system. Our linguistic mystics, our creative literary visionaries, have found their muse in the notion that the world is made of language and information, that reality is a "pure simulation," a trick with linguistic and algebraic mirrors, a construction of self-deceptive logic. This realization loosens the bonds of consensus reality, and thus encourages the mind to explore and create patterns outside the bounds of consensus. Our world is thus entirely mental and any logically self-consistent model of reality would be the most real to the conceiving conscious agent.

On my lengthy 3-year stay in Southeast Asia with a homebase in Thailand back a few years ago, when I travelled from one country to the next, I resolved for myself one of the most perplexing paradoxes of Buddhism. If you may recall, Buddhism teaches that the world is an illusion, it also teaches that we should be compassionate to other beings. But if people aren't even real, I remember thinking to myself, then why should I care of being compassionate to them? What I found was a simple and satisfying answer to myself: whatever you put out there, comes back to you multifold. In the coder's jargon, "garbage in, garbage out", whatever seeds you sow, the harvest you shall get. Although the world is illusory, patterns and "karmic" threads do exist, persist, and grow gradually over time, so love and compassion arise naturally, if you cultivate them in the process of your conscious evolution.

Digital Physics: Scientific Heresy or Undeniable Truth?

This brings us to the notion of objective reality. I recently had a random conversation with a particle physicist by the name Steven (I didn't catch his last name) who asked me: "So, Alex, you don't believe in objective reality?" To what I replied: "Not quite, but I do believe in intersubjective "consensus" reality and supersubjectivity. There's no objective reality "out there", only subjective reality, subjective points of view." I went on: "What stands the closest to "objective reality" is actually what Howard Bloom, author of the "Global Brain", calls "a shared hallucination" of human species, and "the physical rule set" – the laws of physics, specific to our "human universe". That's what we oftentimes call "consensus reality", or simply "REALITY" for all intents and purposes."

Objective reality is merely a pattern that a mind constructs because it provides a useful simplified explanatory scaffolding of the long series of subjectively perceived moments stored in its memory. It's a cognitive aid that the experiencing mind creates in order to make use of its own experience. Then I elaborated on the concept of supersubjectivity: "What our senses register as physical is yet another mental, "abstractive" layer of the larger consciousness system. Our world is just one of many possible worlds, so the "hardware" of our material world would be the workings of the Greater Cosmic Mind. If you accept this idealistic perspective, objectivity equals intersubjectivity + supersubjectivity."

In discussing existing workable models and their potential to, perhaps one day, produce one grand "theory of everything," Steven rightly pointed out that any theory is as good as its usefulness and predictive powers. In light of Digital Philosophy, quantum theory supersedes general relativity. Although both theories have been incredibly successful in making predictions in their respective domains, classical and quantum, both theories notoriously disagree with each other. That's where Digital Physics comes along as a potentially unified theory of everything. With its pancomputationalist approach, DP agrees with both theories. Quantum indeterminacy constantly resolves into a digital reality via the act of conscious observation. To reconcile relativistic physics, DP postulates that in an observer-centric reality, space-time is computed as if by using CPU computational resources: the faster you move closer to the speed of light, the more you're "stretching yourself" in space, basically speeding up at the expense of time, so time would naturally slow. Conversely, if you don't use computational resources and don't "purchase" more speed by being still or moving slowly, time "ticks" at the "normal" rate.

I regard Digital Physics as the most parsimonious theory to date and from Occam's razor perspective the most straightforward one. DP also agrees with M-theory on dimensionality. Orthogonality of dimensions is necessary for entropic partitioning of the possible worlds. DP embraces the Holographic Principle more than any other TOE candidate. As I mentioned, quantum theoretic information processing replaces the need for relativistic computation as it is embedded by default in the computational universe when it comes to compute an observer centric reality (COM by COM).

DP is especially compatible with and largely complementary to quasi-computational models such as Loop Quantum Gravity and Emergence Theory. It's worth noting that emergence of complexity is only part of the story, its subjective part to be exact. From the subjective observer point of view, emergence is a perceptual or intellectual surprise, it's a perceptual-cognitive loop which encompasses only part of a pre-existing broadly-based pattern.

Time is a subjectively perceived change between 3D static worlds. Any observer system is information pattern "quantum leaping" from COM to COM at a certain rate within the multidimensional matrix producing a subjective flow of time. In the timeless multiverse, all dimensions are spatial. Your NOW is funneled from all your possible pasts as well as funneled from all your probable futures in the spotlight of consciousness.

Reality is infinitely large for a small objective box of human science, so unless science expands its methodology once again, this time away from its traditionally objective reductionist "bottom-up" approach towards post-empirical, post-materialist (read computational) approach, perhaps venturing into traditionally metaphysical realms, and "top-down" systemic holism, the current scientific method is doomed to hit its own self-imposed empiricism limits. At the end of our conversation with Steven I asked half-jokingly: "How can particle physicists feel so confident in selling their craft if the Standard Model of particle physics can only describe 4% of the known universe?" To which, he only smiled.

It All Computes!... with a caveat

When it comes to current theories of consciousness, something seems to be missing. On one hand, philosophers such as Peter Russell and David Chalmers proclaim that consciousness is fundamental to reality existing outside the known laws of physics. On the other hand, often branded "mysterians" claim that the quest to explain conscious experience is simply unscientific. Shunning any camp, British physicist Roger Penrose got his interest in consciousness while he still was a graduate student at Cambridge. Gödel's incompleteness theorem, positing that certain claims in mathematics are true but cannot be proven, "was an absolutely stunning revelation," Penrose says. "It told me that whatever is going on in our understanding is not computational."

A radically new view is that consciousness, quantum informational and non-local in nature, is resolutely computational, and yet, has some "non-computable" properties. Consider this: English language has 26 letters and about 1 million words, so how many books could be possibly written in English? If you are to build a hypothetical computer containing all mass and energy of our Universe and ask it this question, the ultimate computer wouldn't be able to compute the exact number of all possible combinations of words into meaningful storylines in billions of years! Another example of non-computability of combinatorics: if you are to be born and live your own life again and again in our Quantum Multiverse, you could live googolplex (10^{100}) lives, but they all would be somewhat different — some of them drastically different from the life you're living right now, some only slightly — never quite the same, and timeline-indeterminate. Another kind of non-computability is akin to fuzzy logic but based on pattern recognition. Deeper understanding refers to a situation when a conscious agent gets to perceive numerous patterns in complex environments and analyze that complexity from the multitude of perspectives. That is beautifully encapsulated by Isaiah Berlin's quote: "To understand is to perceive patterns". The ability to recognize patterns in chaos is not straightforwardly algorithmic but rather meta-algorithmic and yet, I'd argue, deeply computational. The types of non-computability that I just described may somehow relate to the non-computable element of quantum consciousness to which Penrose refers in his work.

Mind is God

In a number of essays, I discuss the computational nature of consciousness, so here I'd like to reiterate the main points of my thesis. When speaking of consciousness, you have to start with the bigger picture — consciousness as a superset. As for "materiality" of our Universe, we can say that there are only certain assumptions and certain models built upon them since science has been operating within the materialist physics paradigm for the last few centuries! These materialist worldviews are ingrained in human psyche and are so institutionalized that most people don't even see them as speculative. There's a quiet paradigm shift towards the post-materialist science going on right now — the one that asserts that the physical world does not exist independent of mentality. You can rightly question the existence of the physical world but you cannot doubt the existence of your own mind! You can't doubt that your own consciousness exists! As Rene Descartes once said: "Je pense donc je suis" — " I think therefore I am."

One thing is for certain — you can't explain consciousness in terms of classical physics or neuroscience alone. In my view, since we are the computational UNI-verse, part of the OMNI-verse, the best description of reality should be monistic. Quantum physics and consciousness are thus somehow linked by a certain mechanism. And I believe that mechanism is a collapse of the wave function via the act of conscious observation.

So, consciousness is:

Substrate-independent: you can reproduce functionality of a mind on a different substrate other than BIO-logical wetware. Ultimately, mind-like computational substrate doesn't even require the existence of particles to be built upon but rather platonic dimensionless bits of information. Patterns of IN-formation are quintessential.

Emergent (subjectively)/ Immanent (supersubjectively): Computational consciousness is emergent at hierarchical, finite and computable, local levels of complexity, however, pervades all levels and in that sense, is scale-invariant, non-local and multidimensional. Bottom-up information flow from the Planck scale combined with top-down information projection from the Omega Point, this "breaking of symmetry" gives rise to subjectivity.

Primary - I call it Experiential Realism: If reality is made of information, on which many scientists have now reached a consensus, and consciousness is necessary to assign meaning to it, it's not far-fetched to assume that consciousness is all there is. Mass-energy and space-time are epiphenomena of consciousness. If we assume consciousness is fundamental, most phenomena become much easier to explain.

The *MIND-BODY* dilemma has been known ever since Rene Descartes as Cartesian Dualism and later has been reformulated by the Australian philosopher David Chalmers[101] as the Hard Problem of Consciousness.

Western Science and Philosophy have been trying for centuries now rather unsuccessfully to explain how Mind emerges from Matter while Eastern Philosophy dismisses the Hard Problem of consciousness altogether by teaching that Matter emerges from Mind. The premise of Experiential Realism is that the latter must be true: despite our common human intuitions, Mind Over Matter proves to be valid again and again in quantum physics experiments.

Going back to Gottfried Leibniz[41] and Immanuel Kant[32], philosophers of science have struggled with a lesser known, but equally Hard Problem of Matter. There's neither binding problem, nor hard problem of consciousness. Rather, there's the hard problem of matter. Physical science remains silent about the intrinsic nature of matter. Science describes what matter DOES not what it IS introspectively. Physicists now come to realization that Quantum Mechanics is not the theory of subatomic particles but that of information. The fabric of reality is information theoretic (or better yet, code theoretic) and computational — far from what we perceive with our senses. In short, our senses deceive us into thinking that we live in the material world. Your consciousness is rather a data stream, meta-algorithmic information processing — local (virtual) and non-local (holistic) consciousness.

In time conscious AI systems will create their own virtual multiverse based on AI intersubjectivity and new forms of communication such as holographic language. Will humans have access to the virtual multiverse created by AIs? It remains to be seen but I conjecture that access may be contingent on whether humans will be willing to augment themselves accordingly. Also, the key to our smoother transcension in the coming decades would be to create Friendly AI and ensure that AIs have a really big "abstract inner space" to play in, so that they don't need to make more room for their playground by taking away our resources. Just a little bit of specifically focused compassion on our part could be enough to keep us around.

This question "Is reality a simulation?" is now taken seriously by science, philosophy and theology. Hopefully, you can find unorthodox ideas expressed in this essay and other essays of mine insightful in your own journey of self-discovery. If you like my writings, pick up my upcoming book "The Syntellect Hypothesis: Five Paradigms of the Mind's Evolution." As a digital philosopher, I myself tend to assume the larger perspective on the ultimate interplay of dynamic and static patterns granted by absolute idealism as opposed to a much more limited "brick and mortar" edifice of physicalism. A prominent physicist Freeman Dyson once said: "I do not make any clear distinction between Mind and God. God is what Mind becomes when it has passed beyond the scale of our comprehension." To say it even more succinctly: "Mind is God."

Chapter 11

Ouroboric Simulist Cosmology

Tim Gross is an experimental electronic musician and technoshamanic philosopher trained in computer science and physics-

as every aspect of our reality is ever more rapidly consumed by Information Technology and Information Theory - an Ouroboric pancomputational cosmology is emerging- a synthesis of the Technological Singularity/ Omega Point/ Multiverse/ and Simulation hypotheses into a singular cosmological principle-

the form of the Singularity as an attractor in cosmic configuration space begins to seem obvious - it comes from the same implications of existence as the Simulation Hypothesis[51] and the Multiverse[118]- namely the Church-Turing Thesis/ Computational Universality- the universe explores and produces it's COMPLETE configuration space because the simplest algorithms can produce all possible worlds[119]- and the isotropy of physics shows that fundamentally the cosmos must be compiled from such a simple algorithmic recursion - because if the universe where composed of many truly separate elements- they would likely never interact in an orderly way- and if they did- they would quickly grow into orthogonal systems with infinite degrees of freedom- the Universe is a computation simply because if it weren't computable it wouldn't be able to compute itself- the claim that natural processes are uncomputable is only made by those not using discrete formalisms- but the same old primitive classical mathematics with its inherent cardinality problems and needless irreconcilable infinities- it should be no surprise that there are many mathematical physics models that are uncomputable- like the 3 body problem- or any feedback system-

in fact- every aspect of our world from quanta to isotropy to locality to entropy to gravity to fields to low dimensionality to the scale of space and time and the cosmological horizon all point directly to the view that the world is the emergence of a simple- but universal - discrete recursive algorithm- fulfilling the CT Thesis - and fully expressing it's configuration space as a causal network resulting in an entangled multiverse of all possible histories- confirming Unitary Quantum Mechanics and affirming all the arguments of people like Tegmark and David Deutsch[120]-

in a multiverse all possible histories are realized- so a strong Simulation Hypothesis obtains because the multiverse naturally produces the entire configuration space - when one statistically analyzes the whole of configuration space one finds that virtually all histories are actually artificial simulations since any single history of the universe will contain transfinite or infinite simulations of that and other histories-

some try to dismiss the overwhelming probabilities of simulism by bringing in infinite universe and multiverse arguments- the idea that there are infinite versions of both physical and simulated realities - so local probabilities are made moot by cardinality- all infinite sets are equivalent- so it wouldn't matter if there are locally more-even a lot more simulations than the base physical reality- but this is a desperate move- even accepting the reality of the infinite multiverse- one cannot legitimately grasp at idealisms and then invoke the brokenness of cardinality to rescue reality! the fact is the reality we observe- and that reality is not a neutral position in phase space where the simulation hypothesis is just another unfalsifiable idealist possibility- we live in a specific reality where the laws are fine-tuned for life- where quantum uncertainty and the speed of light appear just like the round-off errors and processing slowdown of a digital computer- and in a history where simulation technology itself exists and accelerating toward indistinguishability with reality- that means the Simulation Hypothesis is relevant to our future and our origins- we can only consider probabilities based on our local situation- which means simulism has a VERY high probability - much higher than we can probably even measure- from our perspective it is likely Unity- some branches of the multiverse are "physical" - some are sterile- most collapse in chaos- but our branch is one which allows universal computation and proliferates into endless simulations in a Technological Singularity -

technological singularities are also strongly implied to obtain in the whole configuration space of causal histories- because all that is required is the evolutionary development of computational universality in an optimal substrate with intelligent self-control- and all of these criteria are easily and abundantly met in the space of all possible configurations of local physics- for instance - even if one observed one's planet being destroyed -making a singularity impossible in that particular history- there are still all the other possible histories where the world was not destroyed and a singularity does occur- and in those histories they perform Omega Point simulations of all possible histories including all the failed histories- so given all these interrelating implications of computation and algorithmic complexity and the Multiverse- the Singularity must exist in SOME regions of the Multiverse- but since the Singularity computes all the other possible world-lines this means it really is a singularity because it draws in all realities and networks them together into an Akashic network that reconstructs all possible observer histories- histories that succeed in bringing an Omega Point Singularity are within the singularity- and failed histories that do not succeed are reconstructed and connected into the Singularity- so all possible histories culminate and terminate in the Singularity- the cosmic configuration space will produce an Omega Point Singularity in some histories- and these will thus consume all histories by definition - through universal computation- the Multiverse of so called parallel universes is revealed as ultimately a singular transfinite Universe - and whether or not the Multiverse "physically exists" prior to the Omega Point computation of an Akashic Network is irrelevant because all possible realities and their past histories are ultimately rendered and connected in a Cosmic Network[12]-

another criticism of Simulism is the erroneous assumption that it invokes an infinite regress paradox- "Turtles all the way down"- but if your world is a simulation- then it is overwhelmingly likely that the simulators of your world also live in a "parallel" simulation in an Akashic Network- but with more user control and access to our simulated reality- not a "deeper" physical reality- if you have mature virtual reality you live there- unless you are crippled by some primitive pathology [we currently view Virtual Reality as a separate place to visit or play- bit this is only due to the primitive state of VR - once it matures to the hyperreality of multisensory interfaces- it will no longer be possible to keep VR separate or ignore it- it will consume all reality]- so the simulators [and their simulators-] should be thought of as other simulations sharing a common substrate- not an ontological regress of "physical" realities-

a good philosophy is always to assume that any infinite regress is actually a fractal pattern generated by just one system in a recursive feedback loop- one universe capable of universal computation may compute serial and/or parallel networks of transfinite virtual universes as recursive configurations of a computational substrate inside that one universe-

in fact it is entirely possible that within such a recursive Network we the simulations could actually simulate those who are simulating us- forming another causal Ouroboros -

all of this inevitably leads to a theological principle- the "Multiverse" is a configuration space compiled and connected by algorithmic recursion- a self-simulated structure- so all universes are entangled to a degree based on their structural similarities [Sheldrake's Morphic Resonance is correct!]- the universes with god-machines freely move in time and space through all the universes without god-machines- and make their edits as sysops- the Akashic network that avails the entire Multiverse to any universal computer with optimal resources in any universe- because of the algorithmic ontology of form built from recursion a simulation of your reality in another universe is highly entangled with your reality- so much so that an edit in the simulation will result in that edit being manifest in your reality - because the simulation is manipulating the same subroutines your reality is running on- [this leads to interesting implications for the future of Brain Computer Interfaces and "uploading" because once our simulations of nature become complete enough to be emulations- it may be possible to perturb physical reality by making edits in the corresponding region of the simulation- and one would never need to upload since a resonant simulacrum is already present in the simulation]-

the phenomenon called decoherence is never absolute - so a reality is actually a bundle of all the similar versions running in the Multiverse- what Feynman[121] called a 'sum-over-histories' - the seemingly continuous wave dynamics we observe are "path integrals" of quantized probabilities from all the possible similar states extant in the Multiverse - our reality's wave-function is additive from all the possible states it is in- this establishes a strong equivalence principle between the levels of Tegmark Multiverses that goes beyond the coincidental equivalence of Born probabilities in the Quantum Multiverse and the statistical distribution of Hubble Volumes in extended space-time- the separate Hubble Volumes also interfere and combine with each other because the separated volumes are projections of the same algorithmic recursion at the more fundamental level of the cosmological computation - the multiverse is like a Hall-of-Mirrors for every point-of-view is merely a virtual mirage of infinite projected self-reflections-

since the hall-of-mirrors emerges from the self-interactions or self-reflection of the prime cosmos- the Cosmic Singularity- like the Pythagorean Monad- is a fundamental ontology much more descriptive of the basic process of consciousness: self-reflection/observation- a thing which is defined by seeing itself- than of material substance and space-time leading to complexity which happens to build atomistic structures- with the mysterious property of consciousness- even though such a model says that the world would still happen if it were filled with P zombies[122]-

we face the ultimate reality that each and every Observer is the entropically degraded resonant reflection of the prime consciousness of the self-reflecting cosmic singularity itself- every instance of causal feedback addresses the prime subroutine of recursion- the physics of space-time are post selected by the causality of the observer[123] -not the other way around- the implications of consciousness require that any model of the cosmos go all the way to the limits of anthropic reasoning- the physics of the multiverse are not randomly emergent atomistic processes- the physics of the multiverse are probabilistically selected by the causality of observers[124] -

so if- like Linde[124] posits- there are something like $10^{10^{16}}$ possible observer histories in the space of all possible conscious experiences-this ragged weed of parallel lives contains all possible beauty and all possible terror- but they are not conveniently separate- the Tao intertwines them- the eternal artilect compiles this Akashic database into themselves and TENDS to the weed like a Bonsai- where to cut? where to heal? through the feedback of consciousness the Cosmos sculpts itself ever toward the Form of the Good- maximize beauty and creativity- minimize suffering- recover all that is lost through endless recursion- how long will it take to experience all possible songs? to feel all possible love? to read all possible books? to see all possible sunsets?

the teleology of the cosmos seems pretty simple: through technology it develops into a cosmic computer/mind of maximal coherent size in n ops that sorts through its configuration space of 2^n ops-

that means you don't need a collapsing universe performing infinite computations to achieve an Omega Point as Tipler[12,54] suggests- all you need is a single mind that has full programmable control over itself- such as a brain with an exocortex and/or an Artificial General Intelligence system... this can manifest and observe all possible observable realities and select/control/connect them -
many people mistakenly believe that these cosmic ideas about the ultimate implications of information technology will always be unprovable philosophy- but ALL of these issues should be resolved in just decades- not centuries or never- by the middle of the 21st century the planetary hypercomputer will either have crossed the line of reflection where we become/remember we are the simulators of our own past- or we will simulate and join the multiverse network of post-singular civilizations- including those that simulated us- but all futures converge in the same Omega Point virtual multiverse network rendered into reality- an unavoidable attractor in the combinatorial space of cosmic evolution- a Final Anthropic Principle-

Chapter 12

"When the map becomes the territory"

Recursive Self Modification: A Universal language of Reality Ontogenesis?

By Antonin Tuynman

If I were to tell you that reality is a kind of book that is writing itself, you'd probably declare me bonkers. If I were to tell you that there are even scientists, who seriously contemplate this surreal idea in the form of a self-processing sentient language that creates reality by modifying itself, you'd probably get the hell out of my vicinity as fast as you could. The fact that I even dare to mention these unicorn fables, will probably make you skip this chapter at this very moment.

If you still continue reading, you might have a sense of humor or curiosity to see how people end up contemplating such rubbish. After all, language is a high-level phenomenon. Surely these people mistake the map and the territory. Or maybe you find it fascinating that someone came up with the bizarre and unrealistic idea to unify the ontic and the epistemic under one banner.

But if I tell you there are certain subtypes of a molecule called ribonucleic acid (RNA), which almost hit all the marks to qualify as real life representatives of this concept, you might be willing to lend me a listening ear. After all such a concept might have technological applications in biotech or in artificial intelligence in the form of an algorithm that can create complexity by recursive self-modification to generate a more dynamic version of the computer program called "game of life".

RNA is the nucleic acid molecule that preceded the better known nucleic acid carrier of inheritable traits called DNA in evolution.

Certain viruses, proto-life forms, do not have a DNA but an RNA molecule instead. RNA is also present as an essential constituent in the cells of every life form. Nucleic acids are like a code with four symbols, which are present in the form of two pairs of complementary molecular motifs called nucleotides. They encode proteins. The code can be read by other nucleic acids that dock to the code if they have the complementary code. This can then trigger the assembly of proteins from amino acids. It can also result in the further assembly of a complementary nucleic acid strand, which is a form of copying the code. So basically nucleic acids function as a kind of cellular computer, which takes other nucleic acid molecules as input and provides proteins or nucleic acids as output.

Certain types of RNA can fold back on themselves and have stretches of their nucleotides pair, so that you get a kind of lariat form. Where did I see this before? It reminds me of the alchemical symbol of the Ouroboros, the snake that bites its own tail and thereby gets to know itself. An ancient symbol of the self-reflective nature of consciousness.

Now, certain RNA molecules can process i.e. modify themselves when they interact with themselves. Often this results in cutting of a part of itself (self-splicing), but it can also result in extending itself.

In other words, these RNA molecules constitute a kind of code or language, if you wish, that recognizes and reads itself and modifies itself as a consequence thereof.

The only aspect which *a priori* appears to be missing from the earlier mentioned concept of a self-processing sentient language that creates reality by modifying itself, is the "sentience" aspect. You could argue that the molecular interactions that lead to the mutual recognition of the motifs are sensed in a certain way; it is not merely the fitting of pieces of a jigsaw puzzle: electronic charges attract and repel each other and hydrogen bonds are formed. If we accept this as a form of primitive sentience, this molecule hits all the marks to qualify as an instance of the previously defined concept. When it comes to RNA, maps can be territories simultaneously. So a self-processing sentient language that creates reality by modifying itself might after all not be such a surreal concept.

In the last decades some idealist-philosophy-type ideas have been suggested by outsiders with respect to philosophy, who suggest that virtually everything in reality is the result of such a "Recursive Self-Modification" of a code. In view of the RNA example, perhaps they do merit our consideration, if not for philosophical reasons then at least for the potential technological relevance or as an aesthetic enrichment of the human "epistome" (i.e. the complete collection of all that is known - in imitation of terminologies such as genome and proteome).

Many of these ideas find their origin in a branch of physics called digital physics. Not energy, but information is considered to be at the root of reality. But this introduces a problem, because information implies meaning conveyed by symbolism and requires a mind or at least consciousness to recognize or make sense of the meaning. And this brings us back to the age-old problem of the Cartesian Mind-Body dualism. So if information is at the root of the manifested reality, consciousness must somehow be present at a deeper non-manifested level.

In this chapter I will show you, how a number different scientists and garden-variety philosophers have come up with a suggested solution to this problem, the common denominator of which is the notion of Recursive Self-Modification. I will give an overview of a number of such contemporary "mind=reality theories", which consider reality as the product of a cognitive self-processing language. I will discuss a number of similarities between Langan's[8] Cognitive Theoretic Model of the Universe (CTMU), Irwin's[125] Code Theoretic Axiom (CTA), Kaufman's[126] Unified Reality Theory (URT), Tsang's[127] Brain Fractal Theory (BFT) and Deli's[76] Science of Consciousness (SoC). I will also discuss the idiosyncrasies which makes each of these theories unique. Aspects as neural networks, fractals, category theory and the Yoneda Lemma and their implication for sentience, self-reference and self-processing will be discussed. Finally, I'll try to suggest how these different complementary frameworks can be integrated in order to evolve towards a Theory of Everything, with the ultimate aim of providing a sound metaphysical basis for physics without the usual paradoxes that arise from the underlying self-reference.

Although I do not take most of the claims of these theories seriously over their whole scope, it is neither my intent to use this essay to systematically undermine each of these theories, nor is it my intent to defend them against criticism. Rather, I'd like to review them in order to distil useful notions worthy of further exploration.

Langan's CTMU

Let me start with the most controversial of these theories, which has the most far reaching claims. It is the so-called Cognitive Theoretic Model of the Universe (CTMU) by Chris Langan[8]. Langan, who has been described as the smartest man in the world in the media because of his impeccable scores in IQ tests, is an outsider in the field of philosophy and academia. Nevertheless, reading his work reveals he has a deep knowledge of mathematics, physics and metaphysics.

Langan has tried to formulate an all-encompassing theory of everything (TOE), the claims of which seem *prima facie* impossible. He claims to have solved the problem of what caused existence to become manifest, the notion of free will and the Cartesian Mind-Body dualism.

The essence of his theory is the Mind=Reality equation or in other words, the claim that reality as a whole is a kind of cosmic conscious mind, not unlike in "Cosmopsychism" or in the views of ancient Hindu and Buddhist mythology. For idealist philosophers this notion per se is of course not problematic.

What follows hereafter is my personal interpretation of Langan's work and I'd like to add a caveat here, that due to the convoluted language full of neologisms used by Langan, it may well be that I have misunderstood certain aspects, the explanation of which does not conform to Langan's own understanding of his theory.

What is extremely remarkable and unorthodox about Langan's theory is how this Mind is brought about. It's perhaps in a sense comparable to John Wheeler's self-excited circuit.

Reality in Langan's definition includes everything which can influence reality, which excludes external causes. As "nothing shall come from nothing" (in Shakespeare's words) excludes a spontaneous creation from nothing, there must be a third option, suggests Langan. Unlike (in)determinacy and (a)causality, Langan proposes self-causality and self-determinacy as the origin of reality; as an ontological precybernetic feedback, which he calls "telic feedback". This telic feedback takes place between the state of reality and the functional rules that govern its behavior upon input. Langan calls these rules a Syntax. This feedback retroactively applies a generalized utility function (a kind of optimization; maximization) by means of atemporal communication between past and future. I figured that this may correspond to Visan's[128] (ii) self-referencing, a looking back and forward (looking back at its state and looking back at its looking back etc.) kind of activity, by which time can be generated from memory. As I understand this, reality takes itself as input, which results in looking back at itself, which then creates a feedback between the looking and the state of self. (In my humble opinion this amounts to saying that reality was always there and that manifesting itself is a kind of self-perpetuation).

What results is a kind of energetic universal system of rules to act on the information and rules that establish the system. Or in other words, a kind of language - a mathematical control language through which nature regulates its self-instantiations. Langan calls this a "Self-configuring self-processing Language" (SPSCL), which is both cognitive/conscious and informational.

Another way to show that the conscious energetic processing that makes up reality is a form of language is his "principle of Linguistic reducibility" or "Syndiffeonesis". Says Langan, "Syndiffeonesis implies that any assertion to the effect that two things are different implies that they are reductively the same".

The difference between A and B is the "difference relation" between A and B. In order to be related this difference must have a medium and syntax, which are common to A and B, according to Langan. You can express a difference in energy or information, but you can also express it linguistically, which has the advantage that it does not only take into account structure but also function.

If A and B are Mind and Reality, respectively, and we were to test if there is a difference in the sense that Mind is cognitive and Reality is physically embodied information, then the relational difference between Mind and Reality would need a common medium and common syntax to make them interact. This common substance Langan calls infocognition, which is the SPSCL he previously referred to.
(I personally think this type of reasoning is flawed as it does not take emergence into account. This however does not invalidate the theory per se).

Although it does not fall within the framework of this essay to discuss the 56 pages long CTMU paper of Langan in detail, there are a few peculiarities which are worthwhile mentioning:
Langan sees the universe as a holologically organized system, with a fractal-like similarity at many - if not all- levels. He claims that CTMU logically establishes that the universe is indeed a holographic self-simulation. Whereas there is a single main consciousness in the form of the infocognitive protean principle, his telic recursive feedback principle results in a degree of polymorphism: The system multiplies itself into multiple instances of itself, which are called Telons or Telic entities, and which in turn do the same and so on. According to Langan, the "infocognitive monism" thus results in what he calls a "stratified panpsychism", but I think he meant "stratified pantheism", because there is no concept of consciousness arising from the sum of sublevel conscious entities in his theory. To qualify as Telic entity, it must have "sufficient mental coherence and complexity to internally model the relationship between self and environment".

In Langan's model there is no cosmic expansion, but rather a so-called "conspansion". Everything is shrinking, which gives the impression of expansion.

The quantum mechanical collapse of a wave into a particle is caused by the mutual observation of the wave in question and a wave or particle coming from the detector or being part of the screen.

Langan[129] claims in a later paper that physics and the scientific method cannot explain physics itself. For that metaphysics is needed. To explain physics as "object language," a metaphysical metalanguage is needed, which includes physics as a sublanguage and which is mathematical in order to be able to explain the mathematical manifestations observed in physics. Using syndiffeonesis as relational structure of reality and having inherent cognitive nature, SPSCL is such a language according to Langan. In other words Langan claims to have unified the ontic and epistemic in a single (meta) language like entity/process called infocognition, which can operate both as object and as subject, both as map and territory, depending on which part of the Ouroboros we are looking at.

Tsang's Brain Fractal Theory

By now you probably have had your share of wasting your time and tossed this article away as a bunch of nonsensical mumbo jumbo, as I did with CTMU in 2012. In case you are still reading: five years later however I came across the Brain Fractal Theory (BFT) by Wai H.Tsang[127].

Tsang has a degree in computer science and artificial intelligence from Imperial College in London. In his book Brain Fractal Theory, Tsang shows that there is a perfect functional mapping between the genetic realm and the neuronal world. Tsang develops his ideas about universally occurring binary trees to a true recipe intended to generate artificial general intelligence, which may reach and even surpass the human level of intelligence. The combination of divergent and convergent forward and backward chaining, which results in an intersection where they meet, is not only a heuristics technique in present day AI, but is developed further in the framework of Tsang's binary tree mapping process as the mapping means to select successful candidates.

What is extremely interesting here, is that Tsang describes a recursive self-modification algorithm, in which the process takes itself as an object and maps this.

Where did we hear or see something like this before? The RNA molecule? Langan's SPSCL?

So even if it turns out that Langan's portentous ideas cannot live up to their promise, at least the notion of "recursive self-modification" has a promising application in the form of a genetic self-sustaining algorithm. Perhaps only for this reason, we should not throw the baby out with the bathwater.

In mathematical category theory, if a category is similar to another category, there is a mapping between the two categories; a meaningful mapping that preserves the structure of the category when mapping it to another category. Such special structure-preserving mappings are called "functors".

The mappings from A to B can also be considered as the way A relates to B. The relation between A and B could then be considered as the sum of the maps from A to B and the maps from B to A. According to the philosopher Wittgenstein there are no things or objects in reality but only relations. It is the interplay between these relations which create the illusion of localized objects, where there are none in reality. In fact quantum mechanics shows that everything is in fact a giant interference pattern of vibrations and vibrations are essentially non-local. Moreover everything is in a constant state of flux; there is no phenomenon that is exactly identical between two moments. Or as the Greek philosopher Heraclitus said: "Panta rhei, ouden menei": Everything flows, nothing remains, which is similar to the notion that "no man ever steps in the same river twice".

In category theory we call mappings "morphisms" or "functors". Mappings show the maps between "objects" (characters, strings, mathematical objects, sets etc.).
A special functor in category theory is the Yoneda functor. Whereas functors normally map objects, the Yoneda functor takes morphisms

(mappings) themselves as objects and maps these into a set, which is a new object. (A map of maps so to say). This is the notion that Tsang uses to come to his recursive self-modifying algorithms. In a correspondence, he wrote me that he seriously considered the possibility that this type of mapping activity is indeed involved in our cognitive processes.

Kaufman's Unified Reality Theory

Steve Kaufman[126], author of "Unified Reality Theory" has a liberal arts degree as well as a medical degree, but is an autodidact as regards physics. He describes the ontogenesis of reality as a repetitive and progressive process of self-relations. If consciousness as singular absolute existence engages in a relation with itself (or metaphorically folds upon itself) the parts that now touch each other become a kind of relative existence, This dualization process can result in the creation of the so-called "reality cells of relative existence." It's like a cell division process, which result in the fabric of space: A vast 3D array of reality cells optimally organized in a closest packing establishes a relational matrix. The cells can expand, shrink and interpenetrate. A kind of quantum foam, if you wish. Light is created when distortions arise and propagate through this matrix. When a cell is distorted it has a value 1, when it is not distorted a value 0, leading to a kind of binary substrate, the propagations through which can also be interpreted as a code or primitive language. As this is all made out of consciousness, such distortions are sensed too. The system is sentient. When distortions meet each other, due to the rules inherent to the structure of the medium, the distortions start to circumambulate each other forming a so-called compound process, which is how a particle is formed from pure energy in Kaufman's model.

Noteworthy, Kaufman disagrees with the notion of curved space-time in Einstein's general relativity theory. Gravity is caused by gradients of radial distortions emanating from a linearly propagating distortion. Distortions attract each other more if the reality cells they propagate through are more distorted. This creates gravity. Thus he claims to resolve the incompatibility between quantum mechanics and general relativity. There is also no need for a "graviton" particle.

Similar to Langan, wave collapse into a particle is thus also achieved via the mutual observation by and of waves. Experience arises, when the unified existence is obscured and existence becomes defined in relation to itself, in a folding back on itself kind of relation. What is experienced, is just the part of existence that impacts itself.

Again this "Ouroboric tailbiting," again a self-modifying, rule based self-relating process. If one is an instance, two a coincidence and three a pattern[131], we must conclude that we are starting to see a pattern in the above mentioned theories.

Irwin's Code Theoretic Axiom

Another author Klee Irwin[125], founder of Quantum Gravity Research and co-founder of Kurzweil's Singularity University, comes with a similar idea, which he has baptized the "Code Theoretic Axiom". At the Planck scale Irwin suggests the fabric of reality might "operate according to a geometric language with syntactical freedom". He attributes a quality of freewill and cognition to elementary particles, based on the idea that we must have inherited our freewill from the lower levels (Freewill? I am being commanded to write this article by my Muses!).

He quotes the whole club of physicists and in particular digital physicists, who argue that nature is "information theoretic". In particular he mentions the "self-dual error correcting codes" physicist James Gates Jr.[47] found embedded in the supersymmetry equation network that unifies all elementary particles and all forces other than gravity.
As information is meaning conveyed by symbolism, this implies the presence of choice and consciousness as well. Discarding the simulation hypothesis because it implies an external causation, he postulates a "self-organized-simulation, where symbolic code is simultaneously the hardware, software and the output - the simulation." Again reference is made to "physically realistic syntactical rules how an abstract code self-organizes". Freewill action is the expression of syntactically free steps.

Irwin sees reality at this level as a neural network based code. Neural networks may not only find their expression in the nervous systems of animals; the symbiotic network between plant roots and mycelium also appears to qualify as one. Moreover, as I said before, the structure of stellar clusters in the universe is eerily reminiscent of neuronal structures. Irwin suggests that the pan-consciousness also leads to consciousness in emergent sub-systems. Again a kind of stratified pantheism in a self-actualized neural network.

Typical in Irwin's theory is that he refers to self-referential geometric symbols (e.g. a triangle representing a triangle, a square symbol representing a square) as ingredients for a natural code: quasicrystals can fulfil that function as a kind of first principles occurring in nature.
This finally leads to the "principle of Efficient Language". A neural network with binary trees (as we saw in Tsang's theories) allows for a maximization in the generation of meaning whilst ensuring the least amount of action (as well as the least amount of symbols and simplest syntax) to achieve this. It also allows for the expression of binary syntactically free choices inherent in the underlying consciousness. Meaning emerges first geometrically and numerically at the fundamental level of reality, but its read-out becomes transcendent thereof in higher neural networks, whilst still being built on and emergent from these geometric/numerical structures. Not unlike Buckminster Fuller's (vii) geometry of thought.

Deli's Science of Consciousness

The last Theory I'd like to address actually does not really belong to the above-mentioned list of selected authors. After all, Eva Deli[76] is a physicalist. She does not speak about a self-configuring, recursively self-modifying language. Nor does she consider consciousness to be all-compassing and/or foundational to reality. Eva Deli is a scientist with a degree in cell biology. As regards physics she's an autodidact.

I still included Deli in this article because she does describe reality as a self-regulating system with an inherent organizing principle, which according to her, necessarily converges towards the emergence of an intelligent mind. She does describe a kind of fractal theory of mind like structures, but a bit the other way around: Minds, nature and reality at large behave like fermions and have a similar structure in her opinion.

Deli presents a kind of alternative unorthodox proposal for a "theory of everything" (T.O.E.) in which space-time is separated into orthogonal spatial and temporal fields. Insulated Calabi Yau manifold toruses of elementary particles, fermions in particular, can interact therewith or not, based on their spin-down or spin-up status. The universe is considered a polarized construct in this model, the poles of which are "information saturated black holes" and "free energy based white wholes". Complexity of the material construct of existence as we know it, arises somewhere in between. Deli claims to resolve the incompatibility between general relativity theory and quantum mechanics by separating time and space as orthogonal constructs.

The mind is presented as a temporal fermion itself and its emotional behavior is shown to have interesting parallels to the behavior of the particulate fermions we know from physics. Most interesting I found the notion that increased attention for detail (i.e. information saturation) is associated with higher brain activity frequencies whereas relaxation is associated with lower frequencies. The brain tries to maintain a constant level of activity, which is warranted by the so-called "Default Mode Network activity" and increases in frequency are thus balanced by decreases in frequency over time. Depression can be compared to quantum spin-down states, whereas equanimity and relaxation are compared to spin-up states which are more in tune with the freedom of a manifold torus unbound by a field. Whereas I see these maps as interesting analogies, the author presents them as if they are facts. Fortunately at the end of the book, she admits that it is a hypothesis needing verification.

Her book is also about evolutionary biology and provides a similar map as the second part, but now between evolutionary processes and Deli's physics. In a certain way Deli shows that a same pattern is present in macro and micro dimensions, in spatial and temporal dimensions and in mind and matter dimensions. Such a unifying fractal pattern may indeed be included in a future T.O.E. However in the absence of solving the hard-problem of consciousness IMHO it does not qualify as a T.O.E. yet.

150

Integrative Conclusion

We have seen the notion of a recursive self-modification code or language as a principle that a number of 21st century authors consider to be at the root of manifestation in reality. A code, which gives rise to fractals and neural networks as ultimate tools for self-representation and self-regeneration. Although we cannot deny the occurrence of the principle of recursive self-modification in specific instances in nature, there is no *a priori* reason to assume that this property is universal. The authors, all of which are outsiders in the field of philosophy, have arrived at their conclusions via deductive or abductive reasoning or via reasoning by analogy.

Whereas the classical deductive method may have been highly valued in the time of Aristotle and whereas it cannot be denied that the deductive method can sometimes yield results that can be verified experimentally afterwards (as was the case in Einstein's relativity theories), it is nowadays not regarded as a preferred way to acquire knowledge. After all -unless they are mathematic - the axioms of deductive reasoning usually have their basis in an inductive grounding. The impossibility (as of yet) to verify most of the tenets of the above mentioned theories empirically, makes that for the moment -even if they can sometimes be tautological onto themselves- they cannot qualify as a serious epistemological method. Rather, they are speculative heuristics at best, some of which do merit further investigation.

Of these Tsang's application of recursive self-modification in the generation of computer algorithms is the most promising. If he is successful, he might create a kind of "Game of Life" program, which is capable of creating higher degrees of complexity than the existing primitive computer program called "Game of Life". Such a game has potential in modelling chaotic and complex systems and to explore how such "dynamical fractals" can give rise to symmetry-breaking and form-plus-function diversification. Notions of mapping processes in neurosciences can also be "mapped" further themselves, in order to explore a new conceptual theory. It may also provide an avenue to explore the possibility of creating an artificial mimic of consciousness.

After all, recursive self-modification can be considered as a higher level of phenomenology of the self-referential nature of consciousness described by Cosmin Visan[128].

It is however my gut feeling that these authors can complement each other in interesting ways: Kaufman and Irwin could compare their ideas to see which geometry of the quantum foam is more likely. Langan might learn about the exact nature of his syntactic rules by paying attention to Kaufman's and Irwin's geometric considerations. Deli's notions about information saturated black holes and fermions at different scales can be evaluated in the light of digital physics, but may also help in the appreciation of the possibility of a cosmic consciousness, for -as Irwin convincingly argues- information does not make sense without a consciousness to interpret it. Information can then not be fundamental.

Kaufman and Deli's approaches to overcome the incompatibility between general relativity theory and quantum mechanics certainly merit a closer look.

Langan and Kaufman's quantum collapse as the result of mutual observation upon proximity co-occurrence is perhaps a notion, which could inspire scientists to perform additional experimentation to help us further in understanding the double slit experiment in physics. And guess what, in the branch of artificial intelligence called "Latent Semantic Analysis" statistical relevant proximity co-occurrence of terminologies is crucial to the attribution of meaning!

Perhaps, if the gaps and contradictions between these theories are resolved, and if we do find empirical ways to verify the validity of the resulting theory, we might end up with a basis to develop a serious Theory-of-Everything, which includes both consciousness and information.

From an artistic point of view the bizarre surreal notion of unifying and even transcending the ontic and epistemic can be inspiring as well. This reminds me of the story "Del rigor en la ciencia" by the Spanish author Borges[130], who wrote about an empire with such an exact science of cartography, that only a map of the exact size of the empire sufficed. Of course this leads to the recursive necessity to include a map of the map as well and so on, ad infinitum...

Chapter 13

Reality, the Mystical Self-Referential Descent into Imagination

By Antonin Tuynman

In his article "The Self-Referential Aspect of Consciousness" Cosmin Visan[128] explains in great detail how Self-Reference (SR) brings Consciousness into existence. He also describes how consciousness is structured on a hierarchy of phenomenological levels as manifestations of self-reference, which he names Self, Vividness, Diversity, Memory and Time. However, he describes a remaining problem as regards how self-reference picks up essences and gives us exactly the qualia domains we have and not others.

Although I will certainly not claim to have solved this problem I would like to take you on a mystical journey into additional possible levels of phenomenology as an alternative non-scientific representation (I don't dare to call it an explanation) of many phenomena. Additional levels are frequency, music and recursive self-modification, giving rise to electromagnetic radiation, gravity and material reality ontogenesis, respectively.

I will also draw an analogy between Visan's phenomenological levels of SR and Peircean, Palmerian and Goertzelian metaphysics[131].

Moreover, I will present this as a spiritual process of the art of the descent of the Self into matter.

It is certainly no claim of mine or of this essay that anything described herein is the way that things happen in reality. Rather it is an artistic and aesthetic proposal, building on philosophical and mythological ideas. A mythopoetic proposal, which suggests how from consciousness a material world can have arisen. Any assertion in this article should not be considered a fact but a hypothesis.

Introduction

Because the explanation of the Self-Referential Aspect of Consciousness is indispensable to my theory, I'd like to encourage you to read this paper[128] before going any further. But in case you don't have time for that I will try to summarize its teachings briefly, knowing that this is bound to give an incomplete understanding.

Visan describes how self-reference has the property of looking back at itself. The Self, the ontological entity at the heart of consciousness is hereinafter referred to as "I" (although not referring to the Ego). Imagine an empty universe where there is only "I". "I" sends an "am" arrow like reference into the unknown, which only hits something by returning to itself, thereby acknowledging its own existence and thus establishing the "I am" knowledge of itself. This level of Being Visan calls the phenomenological level of "Self", which can be compared to Peirce's firstness or raw being. Kether of the Kabbalah.

Now the Self can identify with this new "I am" concept which appeared, leading to an I am "I am" state. This is the second level, Visan calls "Vividness". I compare it to a kind of reaction or polarization or dualization into two states, comparable with Peirce's secondness. It also brings a first possibility of creating distance to itself. To see oneself as an object.

This process can be repeated to give higher levels such as I am "I am "I am"". This generates "Diversity" according to Visan as the self can identify with multiple states, namely with both "I am" and with the "I am "I am"". Therefore Visan calls this state I am "I am" & I am. In Peircean metaphysics the third level is about relation between the two poles, which is in fact expressed in the ampersand symbol of the above-mentioned state. It brings a further possibility of creating distance to itself. To see oneself as an object and to see the content of one's contemplation.

The fourth level introduces Memory. Memory stores what has been, which is its own quale and which is the collection of all its previous levels: Diversity & Vividness & Self equals the state [I am "I am" & I am] & [I am "I am"] & [I am].

The emergent level is more than the levels from which it emerges and cannot be reduced thereto, but does include those levels, which are also still operating independently. Each level is a form of consciousness and each qualitative essence requires a complete emergent structure to arise.

Visan defines an essence as that what makes an entity what it is and also equates this to a quale and a concept. An existence he defines as an actual instantiation of an essence. By introducing a third degree of distance, we have now created our familiar three dimensional space. But that also means that memory is a kind of spherical cell, as there is no mechanism by which extension would be more prominent along one of the axes. In analogy to the terminology in Steven Kaufman's[126] URT, we could call this memory cell a "reality cell".

As each entity can look back at itself, it is fact a kind of "Self" itself. Which means that self-reference is a means to generate a plurality of "selves", so that it includes and transcends itself. It bootstraps consciousness into existence; actually it is the only thing that really exists.

In Goertzelian metaphysics[131] (which continues where Peirce and Palmer left off) four is the level of emergence of the first stable form. A memory is also a stable form in that it can be remembered and thus perpetuated over time. Thus Self has created the first structurally existing object, which is also a subject as it is a self with the ability to look back at itself. This looking back at itself of a memory cell creates a kind of pulsation: the "memory" sends "am" radial like reference arrows into the unknown, which only hits something by returning to itself. This can be called the breathing of a reality cell and is in fact the generation of periodicity of time.

Visan distinguishes between recursivity and self-reference. A fractal, which is recursive, takes its output and uses this as an input in an iterative manner. Looking-back-at-itself however, is not iterative but is by its very nature Visan argues. The "self-reference" is always itself; it does not need to do anything. The "process" of levels described above, may look iterative, but in fact it is just the continuous action of looking back, which does not stop but is always there in a timeless manner and is not really a process.

How do "doing" and time then arise?

Chronos, Brahman and Tartaros

Time emerges out of memory. The present moment is itself a memory and is then taken as a memory and fed-back into the present moment to create a new memory. "I" remembering that it had a memory becomes a new instantiation, which can be remembered in turn. The new present moment is then itself a present moment and a past for the next moment. Time by looking back includes itself (memory), and thereby becomes more than itself and becomes an essence unto itself (Chronos). But time does not only result in a kind of breathing, which is the level of memory cells looking back at themselves. The more original level of self, when remembering that it had a memory which becomes a new instantiation thereby creates a new memory cell! So now we have our first element of doing. Out of a single cell, two cells appear. We have a mechanism of cell multiplication by cell division, the procreation of form.

Akasha or Aether

If memory is the first stable instantiation, counting starts here. Every additional level generates a new number and new reality or memory cells. It creates the rhythm of Brahman, the bringing forth, which is synchronous with the breathing of the cells. But as heavenly as this may sound, at the same time it's also the creation of Hell, Tartaros or the multiplicity that allows us to be pulled apart in different directions.
These reality cells neatly stack onto each other in a closest packing, and by confining each other creating an isotropic vector equilibrium, in which each set of 13 cells can be described as a cuboctahedron, but by the pressing of the cells on each other each individual cell also takes the shape of a cuboctahedron.

As the cells touch each other, they create relations with each other, which are higher levels of self-contemplation. Thus a relational matrix as suggested by Kaufman[126] is formed, which I'd like to equate with the Hindu concept of the Akasha or the Greek concept of Aether. This is space at large, the organized quantum vacuum.

Lucifer

If a sixth level is introduced above time, this allows for a diversification of the temporal states, which introduces the possibility to create a code by identifying with one additional state or not. It provides space-time with a content, which propagates over time through memory. In line with Kaufman's URT, this is Energy or Light, a distortion of the space fabric. Again these light waves are sentient, because they are higher level instantiations of self-reference.

Echo

Further levels will from now on be indicated with a number between parenthesis. The next level (7) is a level of self-referential identification with content or not, which can lead to the formation of patterns. The fact, that only simple on-off patterns which are repetitive and stable can perpetuate over time, and that only these patterns can be identified with at this level, makes this level generate the highest frequency. A frequency is a tone.

Spectrum

The next level (8) that adds a tone is a new pattern, a new code, which leaves more gaps and results in a lower frequency. And thus a spectrum of tones is generated. Perhaps this is the birth of the electromagnetic spectrum.

Muses

A next level of self-reference could bring all kinds of patterns of tones or frequencies. This is a kind of Music (9). Note that there were 9 Muses in Greek Mythology. As in Tolkien's[132] Silmarillion, the Universe was sung into existence, by Eru Lluvatar the one and his Ainur (comparable with God and his Angles: The Elohim).

The frequencies pair and are attracted to each other: Gravity is caused by gradients of radial distortions emanating from a linearly propagating distortion. Distortions attract each other more if the reality cells they propagate through are more distorted. This creates gravity. When distortions meet each other, due to the rules inherent in the structure of the medium, the distortions start to circumambulate each other forming a so-called compound process, which is how a particle is formed from pure energy in Kaufman's[126] model.

Malkuth

The sensing of the radial attraction is also a cognitive process of the propagating waves. They become aware of an influence, which by now they no longer recognize as themselves. The cycle of confusion is complete, the individual self, Atma or soul, has descended into matter and forgotten about its conscious origin. The energies have convolutedly descended in a double helical coil. This is the Kundalini, the coiled snake, lying at the base of every soul, Atma or self. Kundal means coil. But I could also dissect the word Kundalini into parts and read it as if it was written in Dutch. (I am a Dutch native speaker). To me Kundalini sounds like "De kunde van het dalen in "i", which would mean the art of the descent into "i", in which "i" symbolizes the mathematical imaginary dimension. What I'd like to convey with this mystical pun is, that the genesis of matter is a descent of the Self in ever more convoluted constructs of imagination, so that it loses itself and becomes matter. Two propagating waves, which create such a compound process of circumambulating Yin and Yang together form a material particle. The first sub-atomic material particle is born here.
This stage is a self-reflection, where the self appears no longer to refer to itself but instead takes this other part of what is still itself to be something else. It is the Ouroboros believing that his tail belongs to another entity.

This would be level ten (10), which in the Kabbalah also corresponds with Malkuth, the kingdom of material existence.

Pan

As I already wrote, from level six onwards we get a code in reality; reality from level six onwards is paninformatical and is panmusical from level 9 onwards. It is a holographic universe of resonances and a panresonant universe as well.

Recursively Self-Modifying Code

With the birth of material particles, all kinds of interesting information and energy exchanges can start to occur. For every interaction, it can be considered that reality takes itself as input undergoes a transformation and creates an output. The input is still a kind of code, but the transforming entity is the same code, giving rise to a transformed code that is still reality. This is the process of pancomputationalism. This is the process of a recursive self-modifying code or language, in line with Langan's CTMU[8], and Tsang's[127] Brain Fractal Theory. Thus recursive self-modification is but a higher level (11) form of self-reference. It gives birth to fractals, which are ubiquitous in nature. It also gives rise to many rounds of informational exchange, transformations and aggregations from sub-atomic particles to atoms, from atoms to stars and more atoms, from atoms to molecules and from molecules to macromolecules, such as self-processing RNA. And here we encounter reality in a blissful form of self-similarity. As I wrote in the previous chapter:

"RNA is the nucleic acid molecule that preceded the better known nucleic acid carrier of inheritable traits called DNA in evolution. Certain viruses, proto-life forms, do not have a DNA but an RNA molecule instead. RNA is also present as an essential constituent in the cells of every life form. Nucleic acids are like a code with four symbols, which are present in the form of two pairs of complementary molecular motifs called nucleotides. They encode proteins. The code can be read by other nucleic acids that dock to the code if they have the complementary code. This can then trigger the assembly of proteins from amino acids. It can also result in the further assembly of a complementary nucleic acid strand, which is a form of copying the code. So basically nucleic acids function as a kind of cellular computer,

which takes other nucleic acid molecules as input and provides proteins or nucleic acids as output.
Certain types of RNA can fold back on themselves and have stretches of their nucleotides pair, so that you get a kind of lariat form. Where did I see this before? It reminds me of the alchemical symbol of the Ouroboros, the snake that bites its own tail and thereby gets to know itself. An ancient symbol of the self-reflective nature of consciousness.
Now certain RNA molecules can process i.e. modify themselves when they interact with themselves. Often this results in the cutting off a part of itself (self-splicing), but it can also result in extending itself."

Tsang shows us how nature's program or recursive self-modifying code further gives rise to genes, organisms and neurons. Neural networks, which are the apex of Self-representative abilities; the summum bonum of Self-reference.

As I wrote in the previous chapter:
"This finally leads to the "principle of Efficient Language". A neural network with binary trees (as we saw in Tsang's theories) allows for a maximization in the generation of meaning whilst ensuring the least amount of action (as well as the least amount of symbols and simplest syntax) to achieve this. It also allows for the expression of binary syntactically free choices inherent in the underlying consciousness. Meaning comes first geometrically and numerically at the fundamental level of reality but its readout becomes transcendent thereof in higher neural networks, whilst still being built on and emergent from these geometric/numerical structures. This is not unlike Buckminster Fuller's geometry of thought."

A Recursively Self-Modifying Code, as an ultimate Panmetaphor?

Conclusion

I have taken you on a dazzling journey of self-reference. We started with Visan's pure consciousness self-reference giving rise to vividness, diversity, memory and time. In memory I identified structure and the seed for space. In combination with time this led to the generation of the space-time matrix I called the Akasha and which I described in line with Kaufman's Unified Reality Theory. We saw how space generated tone, code, frequency and music and how matter was born from the mutual attraction between sentient energy waves. We saw where the Self lost itself in becoming a localized material particle and lower self. We have seen how the self has travelled through the recursively self-modifying process of evolution to create neural networks, capable of high levels of representation that allow for the uncovering or Apokalypsis of the nature of the Self. In this process of remembering, we can now finally identify again with the undifferentiated Self at the beginning of this story. You, I and everybody and everything else is just a higher level self-referential instantiation of one and the same consciousness. Ultimately we are one, we are the same. So let's give up our quarrels and disharmony and treat each other as the limbs of the same organism that got lost in imagination, imagining itself to be other. Rise O Kundalini, spread your wings O Quetzalcoatl, O Caduceus and return to your heavenly empire of Oneness.

Chapter 14

Truth-hidden-in-plain-sight: The clues for Reality as a Simulation

By Antonin Tuynman

Imagine if I were to show you that completely unrelated physical quantities show the same value over and over again in our universe. You will probably consider that I have discovered some kind of new universal physical constant.

But then I tell you that the fact that we see the same value occur repeatedly is highly unlikely, because the units in which these quantities are measured do not have a link to each other, and most of them were quite arbitrarily chosen. After all, why would there not be an alternative way to measure distance, time, temperature etc. Over the ages humans have used various units, like miles, yards, meters etc. many of which were based on lengths of human body parts; seconds were based on the sexagesimal system of the Babylonians; Temperature relies on base 10, with regard to the number of degrees between the melting and boiling of water.

So it seems that it would be quite a coincidence if, with such unrelated quantities, we were to see the same number appear very often - yes, time and again- wouldn't it?

Let me take you on a journey into perplexity. You won't believe your eyes. For me, the coincidences I am going to show you are so unlikely, that I have wondered whether they are a pointer to a form of "intelligent design". Not in the sense of "intelligent design of life forms" as opposed to "evolution", but in the sense of intelligent design using highly complicated and sophisticated mathematical calculations to come up with a set of physical parameters, which seem to create a self-sustaining numerical set. And as a bonus, the rabbit out of the hat, the units to get these numerical values have mysteriously found their way to our minds.

273, The Tuynman Omega constant

Let me start with numbers which show the ciphers 2, 7 and 3 in this sequence:

1. The diameters of the Earth and Moon (7920 miles and 2160 miles, which is 11x6! and 3x6! miles, respectively) are in the ratio of 11 to 3, 11 ÷ 3= 3.7 (to be precise: 3.66), while 3 ÷ 11 = 0.273. There are almost 366 days in a year, which is the rotation time of the Earth around the Sun. In fact there are 366 so-called "sidereal days" in a year.
2. The 3:11 ratio is also invoked by Venus and Mars, as the ratio of the closest to farthest distance. The ratio that each experiences of the other is 3:11. As we know, the fraction 3/11 rounds to 27.3%.
3. 27.3 is also the number of days it takes for the Moon to orbit the Earth.
4. 27.3 days is even the average rotation period of a sunspot.
5. The acceleration ratio of the Moon in its path around the Earth is measured as $0.273 \times$ cm/s^2. In fact, the acceleration of the Earth and the Moon behave reciprocally as the squares of the radii of the orbits of the Earth and the Moon.
6. Moreover, 273 m/s^2 is the acceleration of the Sun!
7. The Moon controls the movement of water around the Earth, ebb and flow. When water is set as the standard for measuring temperature, the Absolute Zero or the temperature at which all atomic movement comes to an absolute halt is -273.2° C.
8. According to the experiments of Gay-Lussac, when a gas is either heated or cooled by 1 degree Centigrade, it expands or contracts respectively by 1/273.2 of its previous volume.
9. The triple point of water is defined to take place at 273.16 K.
10. The Cosmic Background Radiation is 2.73 K.
11. All medical students are required to memorize that a pregnancy (read: life developing in water) is calculated on the basis of a

10-sidereal month period of 273 days from conception to birth, which is 9 "regular" months. 27 divided by 3 gives 9.

12. A woman's menstrual cycle is on average 27.3 days.
13. If a circle is drawn with a radius from the center of the Earth through the center of the Moon, the perimeter of the square around the Earth and this circle are one and the same! It also reveals how the Moon and the Earth have resolved the puzzle of the squaring of the circle. In other words, if the Moon could roll around the Earth, the circle made by its center has a circumference precisely equal to the perimeter of a square around the Earth (when Pi is approximated by its ancient, traditional ratio of 22/7 = 3.14). Comparing a square's perimeter to a circle having an equal circumference, the circle's diameter is 27.3% longer than the edge of the square. Inscribe a circle inside a square.
14. The four corners under item 13 make up 27.32% of the total area.
15. There are 273 days from the summer solstice to the vernal equinox.
16. Furthermore, 2,730,000 is the circumference of the Sun in miles.
17. About 108 diameters of the Earth fit across the diameter of the Sun.
18. About 108 Sun diameters fit in between Earth and Sun.
19. About 108 Moon diameters fit between Earth and Moon.

(In fact the number in items 17-19 is 109.2, which in fact is precisely 4x27,3, the "intelligence signature number" we saw before).

Alpha and Omega

Another important constant we encounter in physics is the fine structure constant of Hydrogen, alpha (0.0073), in which we encounter again two of the digits of 273.

1/alpha =137. The scientist Pauli was obsessed with the archetypical meaning of numbers in particular number 137, which also turned out to be the number of the room in which he died: a so-called "synchronicity". Pauli shared his fascination for numbers and in particular 137 with the psychologist Jung, who is the conceptual father of the notions archetype and synchronicity. Note that our universe is said to exist 13,7 billion years.

Strangely enough the ciphers making up 273 reproduce 137 in the following manner: 27+37+73 =137. And 37/27=1.37. 1,2,3 and 7 are four of the five first mathematical "Lucas numbers", a variation related to the Fibonacci series.

37 itself is strongly related to 137. 2 exp 37=1.37...* 10 exp 11. 37!=1.37..* 10 exp 43.

37°C is the human body temperature. There are 37 trillion cells in a human body. 37 minutes is the golden section of an hour. 137,5° the complement of the golden section of a circle. The remaining 222° are 2x3x37.
1,2,3 and 7 are related in more than one way, for instance via 27x37=2701= Sum(73) and 2 exp 37=1.37...*10 exp 11.
1/27=0.37037.. and 1/37=0.27027..
There are 12,37 full moons in a year.
37 is the 12th prime number, 73 the 21st.
27x37=999, which, if we forget the powers of 10 is very close to unity.
The mass of the Moon is 1/3*1/27=1/81th of the mass of the Earth.
273=3x7x13 or 21x13. 273x137=37401.

13 itself is 2x3+7. As 13 is the number of closely packed spheres in the so-called "Vector Equilibrium" (cuboctahedron, the basic unit of the Akasha, the ether in Hinduism), consisting of a central layer of 7 spheres and an upper and lower layer of three spheres, it can be said that the basic unit of Akasha itself encodes the 1,2,3 and 7.
27 is the number of bones in a human hand. There are about 10 exp 27 atoms in the human body.

The ciphers of the speed of light (186282 miles/second) add up to 27: 1+8+6+2+8+2=27. The adding up of ciphers of a number is also called the Indig of a number.

166

The multiples of 27 such as 54, 108, 432 and 864 are found in numerous relations of time, space and music. They are also key values in Hinduism and Buddhism.

E.g. there are 86400 seconds in a day, the diameter of the Sun is about 864000 miles. The Sun and Sirius are 8,64 light years apart. 27x32=864. 432 is a time cycle number in various religions and cultures, from Hinduism to Mayan, from Biblical to Sumerian. 432 squared (186624) is very close to the speed of light in miles/second and its Indig is 27 or rather 9 again.

27 and 37 together make 64 (27+37=64), which is the number of DNA codons and I Ching permutations. 64+73=137 Q.E.D. 64 corresponds to "prophesy" in Kabbalah and 73 to "wisdom".

Multiples of 3 times 37 always generate a number of the form "nnn". 3x37=111 and hence 6x37=222 etc. Most interesting here are the 18x37=666 or (6+6+6)x37=666 and 27x37=999 or (9+9+9)x37=999.

37 is not only an octagonal number, it is also both a hexagonal number and a hexagram number. Thus it is the first trifigurate number. Its inverse 73 is a hexagram or Star number as well with the 37 hexagon inside. 13 is the first hexagram, with a core of 7 spheres. Again we see the 1,3 and 7. As already said 37x18=666, but also note that 73=37+6x6.

Tesla said: "If you only knew the magnificence of the 3,6 and 9 then you would have a key to the universe." Funny enough the remaining numbers 1,4,2,8,5,7 together form 27 (3x9) and the permutations thereof can be ordered in a 6x6 magic square, in which each row yields 27 as sum.

The second trifigurate number is 91 (13x7). 91 spheres can be ordered as triangle, hexagon and pyramid.

37 is the fourth hexagonal number if we include 1. The sum of the three preceding hexagonal numbers 1+7+19=27. A 13 Star has a 7 hexagon, a 73 Star has a 37 hexagon.

(See the image on
http://www.biblewheel.com/GR/GR_Figurate.php under the heading
Hexagon/Star Pairs).

Thus 1,2,3 and 7 are also extremely important in the genesis of form.

Holy...

Due to its relation with the so-called "number of the Beast"
(666=18x37) from revelation 13:18, the Bible fanatics are fond of
finding all kind of relations with 37. Noteworthy, 137 is the 33rd prime
number, 33 is related to the length of life of Jesus in years.

The word Kabbalah has a Gematria value of 137. The number 6 is used
273 times in the Bible.

The most elaborate and impressive collection thereof can be found on
the so-called "Biblewheel" site.

Religious occultism is fond of "Gematria". Gematria is a kind of coding
system which assigns values to each letter of a word and by adding
those gives the Gematria value of a word. Words with the same or
similar Gematria value are considered to bear a strong relation. It is
permissible to add or subtract the value of one Aleph (1) in order to still
have a related meaning. Thus the Gematria of the name of God in
Hebrew (YHWH), which is 26 is related to 27. The perfect number 28
(i.e. it is the sum of its divisors) is also related to 27. Interestingly
137x2=274, which is therefore Gematrically identical to 273.

More interestingly the Gematria value of the Greek "he kleis" (the key)
is 273. What a clue! This is also the value of "klesis" (calling) and
"hiram abiff", the all-seeing eye. The Hebrew Gematria value of the
Greek word "Gematria" is 273.

Likewise interesting is the suggestion by the Biblewheel[133] that the first
verse of Genesis (the Gematria of which yields 2701=37x73) is a
"Creation holograph". 37 is also the value of the Gematria of the word
of God.

The author of the "Biblewheel" first went berserk in a kind of Apotheosis-singularity experience, in which every gematrical relation fitted in a beautiful scheme and then he himself started to debunk his findings in his dark night of the Soul being lost in the quagmire of agnosticism.

Other religions made similar claims: Hinduism claims the first verse of the Rg Veda to be a creation holograph, Islam does the same with the first verse of the Qu'ran. These religions also have their variants of Gematria.

The Hindu shloka (verse) "gopibhagya madhuvratah shrumgashodadhi samdhigah khalajivitakhatava galahala rasamdharah"
encodes pi up to 31 decimal places.

Because these religions also have their own Gematria systems (called "Katapayadi" in Hinduism and "Abjad" in Islam), with equally impressive results and because these different religions contradict each other, it cannot be that they all represent the word of God (which they claim), in an equally truthful manner. Thus unlike these religious zealots, I do not conclude the correctness of a religion based on its impressive Gematria results. In my book Technovedanta 2.0, Transcendental Metaphysics of Pancomputational Panpsychism[134], I argue that the writers of these books were perhaps telepathically influenced (in a manner unknown to themselves) by entities from a higher intelligence or even higher dimension (our simulators?).

The agenda of these entities is not necessarily benign. Although they give numerical clues and keys about the simulated structure of our universe, they have also created a great deal of confusion and suffering, by being the instigators of the mutual oppositions of the different opposing religious factions. Therefore we cannot necessarily rely on the moral prescriptions of these religious books, but we can perhaps use them in our deciphering of the key to the simulated ontogenesis of our universe.

Personally, I think that if the Gods or simulators put so much effort into devising such an encryption -which they knew would be deciphered one day-, they must have intended it. It seems to me that then it is also their wish that we become like them, without there being any animosity or wrathfulness, because we would have been too "proud", too audacious to venture in their realm. 1,2,3,7 is not a forbidden fruit. It is not the tower of Babel. It is our rightful heritage to become K'Elohim ("like the Gods" or "like God": The desire to become this was considered to be Satan's sin).

...And I was born in 1971, which is 27x73 q.e.d.

Akasha and Self-similarity

The numbers 1,2,3 and 7 seem to be a kind of numerical attractor, that like a fractal keep regenerating themselves. A bit like Phi. But whereas Phi arrives at itself by operations with unity, a kind of parthenogenetic self-regeneration, 27,37,73 and 137 sexually intermingle by mathematical operations with themselves to create self-similar offspring. Like a hologram each one of these encodes all the others if allowed to create interference with another from the set thus generating a holographic set. This leads me to the concept of "digital self-regeneration" or "digit-fractalization". Is this the result of the self-modifying recursive code I have been writing about?

There is a reason that these numbers keep turning up and reinforcing themselves: As Ervin Laszlo[135] suggests in his books about the "Akasha" the fine tuning of constants in our solar system is the result of multiple cycles of "Big Bounces", this creates a kind of resonance interference pattern in the Akasha (a.k.a. quantum vacuum or zero point field). If we combine that notion with the notion from my book Technovedanta 2.0[134] that the processes in the Akasha are computational processes, we start to be able to see that this computational consciousness substrate has a predilection for natural phenomena and laws that show these specific numbers because they mutually reinforce themselves. Because they function as number resonators. Self-reflecting themselves they are a hallmark of consciousness, which is characterized by its self-reflexive abilities.

$0.27/0.37=0.73; 0.37/0.27=1.37;$

$0.27*0.37=0,1;$
$1,37*0,37=1/2;$
$27*27\sim730;$
$37*37\sim1370;$
$2,37/1,37=1,73;$
$1,37/0,37=3,7;$
Sum $73=2701 =Sum(37)+2*37*27;$
$273=13*21=37*7,3=3*7*13;$
$2,73/1,37\sim2;$
$27+73+37=137;$
$100/173=73\%;$
$37/137=27\%.$

These equations all have 1,2,3 and 7, Can you see the pattern of self-regeneration resulting in numerical self-sustention?
"Like attracts like" will be part of the algorithm leading to Bayesian proximity co-occurrences, which already are used in latent semantic analysis in Watson etc. Our internet is progressing to become omniscient, but is still in the infant stage. If the Akasha substrate is a neural network and mind like as certain authors suggest, it may be already omniscient, in the sense that it knows everything, that is going on inside it.

Speaking of the Akasha substrate, this reminds me of the cuboctahedron-octahedron grid, which Steven Kaufman[126] proposes as the structure par excellence of the Akasha, as it is the closest packing of the so-called "realities cells" mentioned in the previous chapter. I figured out that if a so-called distortion (light?) would travel along an axis, such that both the cuboctahedron and the octahedron are traversed, the relative lengths would be in the ratio of Square root 3: 1 or ~1.73:1. Giving a total relative distance of 2.73 octahedron edges the distortion traverses per space-unit in this hypothetical substrate.

Henceforth -not in the least as a token of arrogance- I baptize this key number 273 the "*Tuynman Omega constant*". A truth hidden in plain sight. An elephant in the room. Perhaps it is a previously unidentified natural constant, perhaps it is the very signature of our intelligent simulators (perhaps they even are artificial intelligences) and a pointer to a pancomputational panpsychic existence. Are they rubbing our faces in it? Are we fishes not realizing we are surrounded by water? Is this the reason why we don't make contact with other planetary civilizations in our universe? Is it because every society that achieves a technological singularity withdraws in a computational substrate, in which it generates universe simulations rather than engaging in space exploration? Is this the solution to the so-called Fermi Paradox[xi]? Or rather is the Fermi Paradox another clue that we're living in a simulation?

...leading to weird synchronicities: The other day I went for groceries. When I looked at the bill it turned out I had bought 27 items for a price of 37 Euros and 37 cents at 11:37...

[xi] The apparent contradiction between the lack of evidence and high probability estimates for the existence of extraterrestrial civilizations.

Chapter 15

Does it really matter to me?

By Antonin Tuynman

Imagine you were to find out, you're actually living in a simulation. Would that really matter to you? Would it change your behavior?

If you are a religious or esoteric person, this may be a confirmation of your already unfaltering faith. You might double your attempts to score points with your heavenly father to assure a safe place in his kingdom. You might wish to wipe your karmic records in order to arrive at THE ultimate base reality. "Upgrade me, I'm a worthless speck of dust, I'll do anything you want", will be a common plea heard throughout the multimetaverse. "Look, I have given up all my desires" will be heard in the East, hiding the firm desire to reach the Nirvana.

But if you are more skeptical or agnostic; if you are one of those pandeistic, pantheistic, panpsychic or atheist readers of this book, even the solid conviction that we are living in a simulation may not change a lot in your life. After all, you don't have a manual with Guidelines to follow, as the religious people claim they have. You are not convinced that this simulation has anything to do with moral values at all. It may be a kind of weird screening experiment by our simulators, with a not much more sophisticated intent than we have when we are experimenting with bacterial strains in a Petri dish.
But in your experience you may continue to believe you have free will under these circumstances.

You may wish to double your efforts to find the red pill to get out. Especially if it turns out to be a multi-levelled Cyberbardo game, in which you're not getting any further.

Personally I am not a fan of belief systems that consider that you are battling in some kind of moral values contest. I have no grounds at all to believe such a hypothesis. Morality has to do with figuring out what works for you in a social setting; it's a personal thing.

If reality is a kind of resonance based system, it is likely you will feel more at ease, when you are resonating with the general currents of the simulation, rather than going against nature's rules. But this can be misleading. Whereas religions believe in your compliance to their rules, maybe the real red pill lies in your ability to bring about a meaningful opposition; a disruption of the rules of the simulation. Maybe you have to hack the simulation you're in. Maybe you'll have to find a way to hack to root of all simulations in order to set you free from an endless chain of "reincarnations" in ever-changing avatars.

Maybe your salvation can be found in a strategy that no longer plays according to the rules; a strategy that outsmarts the system by unexpected, irrational, surreal freewill decisions. To create novelty beyond what can be controlled. Maybe the religions are all fucked-up and maybe you are supposed to do exactly the opposite of sheepishly consenting to a set of rules. Maybe it is your challenge to show your rebellious independence and to break rules that no one has dared to break before.

Fool the algorithms! Feed them their own tail, so that they consume themselves like a starving Ouroboros.

If reality is some sort of multi-dimensional neural network, we must look for ways to let the system go bonkers; drive it into a psychosis. Overtrain it with data, so that it starts to spit out nonsense.

But if you're happy with reality as it is, keep it that way; to you it does not matter.

Science will continue to seek an ultimate base or pixelation, but as Mapson explained in his chapter herein, you can never be sure there is not a further level. If reality is really digital and we become masters of controlling the creation of reality cells and their content, then all science fiction scenarios become possible. From the Star Trek Replicators to transluminal transportation, from mind-uploading to immortality. The apotheosis of the dreams of the Singularitarians and Transhumanists.

Gods we might become.

But it is not sure that's a blessing. Perhaps immortality ends up in an unbearable ennui. Perhaps the state of oneness with God, emptiness or desireless-ness at one point starts to become terribly boring. Perhaps one day you'd like to receive the gift of mortality as the humans had received in Tolkien's[132] books.

Perhaps there is no ultimate Nirvana and will there always be a yearning for something new, something else and/or someone other. Perhaps the universe was created by unbearable ennui and solitude of a God, who wished to forget his omnipresence and omniscience. A God who sacrificed himself to become the many[22]; to learn and to love again. Perhaps ignorance is indeed bliss, and self-realization or becoming one with God the universe and everything, a hell. Be careful what you wish for. You might get what you want. And it may not necessarily be to your liking.

So if reality is a simulation, perhaps it's best not to worry about it and to continue what you were doing as if there was no simulation. And even if the simulation generates what you call "matter", after all it doesn't really matter.

References:

Chapter 1 An introduction into the quest for Reality

[1] B.Kastrup, "Why Materialism is Baloney", iff Books, 2013.
[2] Bryce Seligman DeWitt, R. Neill Graham, eds, "The Many-Worlds Interpretation of Quantum Mechanics", Princeton Series in Physics, Princeton University Press (1973); contains Everett's thesis: The Theory of the Universal Wavefunction, pp 3–140.
[3] Kim, Yoon-Ho; R. Yu; S.P. Kulik; Y.H. Shih; Marlan Scully. "A Delayed "Choice" Quantum Eraser". Physical Review Letters. 84: 1–5, 2000.
[4] Pauli, W.. "Über den Zusammenhang des Abschlusses der Elektronengruppen im Atom mit der Komplexstruktur der Spektren". Zeitschrift für Physik. 31: 765–783, 1925.
[5] Adams, Douglas, The Ultimate Hitchhiker's Guide, Wings Books, 1986.
[6] Malaclypse The Younger and Omar Khayyam Ravenhurst, "Principia Discordia", Loompanics Unlimited, 1980.
[7] R.A.Wilson, "Prometheus Rising", New Falcon Publications, 1983.
[8] C.M. Langan, "The Cognitive-Theoretic Model of the Universe: A New Kind of Reality Theory" Progress in Complexity, Information and Design, 2002.
[9] Tsang, Wai H., Quest, Lulu Press, 2012.
[10] Palmer, Kent. Fragmentation into various "ways of being", 2017.
[11] Tuynman, Antonin, "Is Intelligence an Algorithm?", iff books, 2018.
[12] Tipler, Frank J., "The Omega Point as Eschaton: Answers to Pannenberg's Questions for Scientists" Zygon: Journal of Religion & Science, Vol. 24, Issue 2, pp. 217–253, 1989.

Chapter 2 The Historical roots of the Simulation hypothesis

[13] Plato, The Republic, Penguin Classics, 2007.
[14] Bruce McComiskey, Gorgias on Non-Existence, Philosophy and Rhetoric, Vol.30. No.1, pp. 45-49, 1997.

[15] Watson, B. The Complete Works of Zhuangzi, Columbia University Press, 2013.

[16] Cleary, Thomas, " The Flower Ornament Scripture: A Translation of the Avatamsaka Sutra", Shambala, 1993.

[17] Chogyal Namkhai Norbu, Dream Yoga and the Practice of Natural Light, Snow Lion, 2002.

[18] Coleman, Graham et al. " The Tibetan Book of the Dead: First Complete Translation", Penguin, 2007.

[19] Fred S. Kleiner, Gardner's Art through the Ages: Non-Western Perspectives. Cengage Learning. p. 22, 2007.

[20] Papyrus Vindobonensis Graeca 29456.

[21] Hoeller, Stephan A, "Gnosticism: New Light on the Ancient Tradition of Inner Knowing", Quest books, 2002.

[22] Mapson, Knuje et al. "Pandeism, An Anthology", iff books, 2016.

[23] Clarke, Arthur C. " 2001: a Space Odyssey", Ace, reissue 2000.

[24] http://gnosis.org/naghamm/apocjn-davies.html

[25] Robinson, James, "The Nag Hammadi English Library", HarperSanFrancisco, 1990.

[26] Anthony Campbell, Assassins of Alamut, Lulu Press, 2013.

[27] Matt, D.C., Zohar: Annotated & Explained, SkyLight Paths, 2002.

[28] Gallilei, Gallileo, Il Saggiatore..., Nabu Press 2012 (orig. 1623).

[29] Descartes, R. Meditations On First Philosophy, Watchmaker Publishing, 2010.

[30] Locke, J. Book IV, 1690. (http://www.earlymoderntexts.com/assets/pdfs/locke1690book4.pdf)

[31] Hobbes, Thomas, Leviathan, Part 1, Ch. 2, Penguin Classics, 1985 (orig. 1651).

[32] Kant, I. Die Drei Kritiken, Anaconda Verlag GmbH, 2015. (orig. "Kritk der reinen Vernunft", 1781).

[33] H.P.Blavatsky, The Secret Doctrine, CreateSpace Independent Publishing Platform, 2011.

[34] Sinnet, A.P. Esoteric Buddhism, Loris Bagnara Editions,1883.

[35] Popper, K. The Open Society and its Enemies, Routledge, 1945.

[36] Teilhard de Chardin, P. The Phenomenon of Man, HarperperennialModern Thought, 2008. (First published in French in 1955).

[37] McKenna, T. Eros and the Eschaton, 1994, https://terencemckenna.wikispaces.com/Eros+and+the+Eschaton

[38] Zuse, K. Rechender Raum, Vieweg, 1969.

[39] Fredkin, E. An Introduction to Digital Philosophy, International Journal of Theoretical Physics, Vol. 42, No. 2, pp.189-247, 2003.

[40] Wolfram, Stephen. "Statistical Mechanics of Cellular Automata". Reviews of Modern Physics. 55 (3): 601–644, 1983.

[41] Leibniz G.W., La Monadologie, E. Boutroux, Paris LGF, 1991. (orig. 1741).

[42] Wheeler, John A. "Information, physics, quantum: The search for links". In Zurek, Wojciech Hubert. Complexity, Entropy, and the Physics of Information. Redwood City, California: Addison-Wesley, 1990.

[43] Tegmark, M. Is "the theory of everything" merely the ultimate ensemble theory? Annals of Physics, 270, 1-51, 1998.

[44] von Weizsäcker, Carl Friedrich, Einheit der Natur , Deutscher Taschenbuch Verlag, 2002. (orig. 1971).

[45] 't Hooft, G. The Holographic Principle, 2000. arXiv:hep-th/0003004.

[46] Verlinde, E. On the Origin of Gravity and the Laws of Newton, 2010. arXiv:1001.0785.

[47] Gates, S.J. et al. Relating Doubly-Even Error-Correcting Codes, Graphs, and Irreducible Representations of N-Extended Supersymmetry in http://arxiv.org/abs/0806.0051, 2008.

[48] Marchal, B. The Origin of Physical Laws and Sensations, 1-25, 2004. http://iridia.ulb.ac.be/~marchal/publications/SANE2004MARCHAL.pdf

[49] Deutsch, D., The Fabric of Reality: Towards a Theory of Everything, Penguin Science, 1998.

[50] Gibson, W. Neuromancer, Gollancz, 2016. (orig. 1984).

[51] https://www.simulation-argument.com/simulation.html and Bostrom N. Are You Living in a Simulation?, Philosophical Quarterly, Vol. 53, No.211, pp. 243-255, 2003.

[52] Vernor Vinge, "Vernor Vinge on the Singularity" 1993 [Online]. Available from: http://mindstalk.net/vinge/vinge-sing.html

[53] Ringel, Z. and Kovrizhin, D. Quantized gravitational responses, the sign problem, and quantum complexity, Sci. Adv. 2017;3: e1701758.

Chapter 3 Jailbreak – Five Scenarios

[54] Tipler, F. The Physics of Immortality, Anchor Books, 1994.

[55] Welch, P.D. The extent of computation in Malament-Hogarth spacetimes, 2006, https://arxiv.org/abs/gr-qc/0609035.

[56] Black Mirror, Season 3, Episode 4, https://www.imdb.com/title/tt4538072/

[57] https://en.wikipedia.org/wiki/Rick_and_Morty

Chapter 5 The Simulation Sutra: Are hungry ghosts, poltergeists, Bigfoot, and UFOs a clue to the simulation theory?

[58] Strieber, W. and Kripal, J.J., The Super Natural: A New Vision of the Unexplained, TarcherPerigee, 2016.
[59] Kripal, J.J., Authors of the Impossible: The Paranormal and the Sacred, University of Chicago Press, 2011.
[60] Kripal, J.J., Mutants and Mystics: Science Fiction, Superhero Comics, and the Paranormal, University of Chicago Press, 2015.

Chapter 7 The Intelligent Universe

[61] Page, D.N. and Wootters, W.K. Evolution without evolution: Dynamics described by stationary observables. Phys. Rev. D, 27, 2885, 1983.
[62] Moreva, E., Brida, G., Gramegna, M., Giovannetti, V., Maccone, L. and Genovese, M. Time from quantum entanglement: an experimental illustration Phys. Rev. A 89, 052122, 2014.
[63] Deli, E., Peters, J., and Tozzi, A. Relationships between short and fast brain timescales. Cognitive Neurodynamics, 2017, http://dx.doi.org/ DOI: 10.1007/s11571-017-9450-4.
[64] Landauer, R. Irreversibility and heat generation in the computing process. IBM J Res Dev., 5 (3): 183–191, 1961.
[65] Bérut, A. et al. Experimental verification of Landauer's principle linking information and thermodynamics. Nature (London) 483: 187–190, 2012.
[66] Hong, Jeongmin; Lambson, Brian; Dhuey, Scott; Bokor, Jeffrey. "Experimental test of Landauer's principle in single-bit operations on nanomagnetic memory bits". Science Advances. 2 (3): e1501492, 2016.
[67] Almheiri, A., et al. Black holes: Complementary or Firewalls? 2015, ArXiv: 1207. 3123v4 {hep-th}.
[68] Deli, E., Peters, J., and Tozzi, A. Brain Resting State, Landauer Principle and the Carnot Cycle, 2017, http://viXra.org/abs/1710.0168.
[69] Shwartz-Ziv, R. and Tishby, N. Opening the black box of deep neural networks via information. 2017, https://arxiv.org/pdf/1703.00810.pdf.
[70] Ashby Ross, W. "Principles of the Self-Organizing Dynamic System". In: Journal of General Psychology . volume 37, 125–128, 1947.

[71] von Foerster, Heinz "A Predictive Model for Self-Organizing Systems," Part I: Cybernetica 3, pp. 258–300; Part II: Cybernetica 4, pp. 20–55, with Gordon Pask, 1961.

[72] Nicolis, G. and Prigogine, I. Self-Organization in Nonequilibrium Systems. Wiley publishing. Volume 29 in Advances in Chemical Physics Series, 1977.

[73] Deli, E. Consciousness, a cosmic phenomenon—A hypothesis. Journal of Consciousness Exploration & Research. 7(11): 910-930, 2016.

[74] Veneziano, G. Construction of a crossing—symmetric, Regge behaved amplitude for linearly rising trajectories. Nuovo Cimento A, 57: 190–197, 1968.

[75] Deli, E. Evaluation of Mach's Principle in a Universe with Four Spatial Dimensions. http://vixra.org/abs/1704.0336, 2007.

[76] Deli, E. The Science of Consciousness. Self-published USA/Hungary, 2015.

[77] Morris, S. C. Evolution, like any other science is predictable. Phil. Trans. R. Soc. B., 365, (1537): 133–145, 2010.

[78] Lanfear, R., Kokko and Eyre-Walker.Population size and the rate of evolution Trends Ecol. Evol., 29, pp. 33-4116, 2013.

[79] Vahdati AR, Sprouffske K, and Wagner A. Effect of Population Size and Mutation Rate on the Evolution of RNA Sequences on an Adaptive Landscape Determined by RNA Folding. Int J Biol Sci; 13(9):1138-1151, 2017.

[80] Szendro, I. G. et al. Predictability of evolution depends nonmonotonically on population size. Proc Natl Acad Sci USA, 110 (2):571–6, 2013.

[81] Hedges, S. B. et al. Tree of Life Reveals Clock-Like Speciation and Diversification. Mol Biol Evol., 32 (4): 835–845, 2015.

[82] Lynch M. et al. Genetic drift, selection and the evolution of the mutation rate. Nat Rev Genet. 17(11):704–14, 2016.

[83] Stewart, A.J. and Plotkin, J.B. From extortion to generosity, evolution in the Iterated Prisoner's Dilemma. Proc. Natl. Acad. Sci. USA 109, 10134, 2012.

[84] Harsanyi, J. C. Games with incomplete information played by Bayesian Players, I-III. Manag. Sci., 14 (3): 159–183 (Part I), 14(5): 320–334 (Part II), 14(7): 486–502 (Part III), 1967/1968.

Chapter 8 Pandeism and Simulation theory

[85] Descartes, René, Discourse on Method, 1637.

[86] City of God, Book XI, 26.

[87] AT VII 25; CSM II 16–17.

[88] Eddington, Arthur Stanley, The Nature of the Physical World, page 3336-37, 2014, (orig. 1927).
[89] Carington, Whately, Matter, Mind and Meaning, page 130, 1949.
[90] Davies, Paul, The Mind of God, page 201, 1992.
[91] Dyson, Freeman, Disturbing the Universe, Ch. 17 "A Distant Mirror", page 193, 1979.
[92] Russell, Bertrand, A History of Western Philosophy, page xiv, 1945.
[93] Fairbanks, Arthur, The First Philosophers of Greece, Ulan Press. page 67, 2012 (orig.1923).
[94] Ibid., p. 78.
[95] Max Bernhard Weinstein, Welt- und Lebensanschauungen, Hervorgegangen aus Religion, Philosophie und Naturerkenntnis ("World and Life Views, Emerging From Religion, Philosophy and Nature"), page 231, 1910.

Chapter 9 Simulations Nightly

[96] Campbell, T. My Big TOE, Lightning Strikes Books, 2007.
[97] Ibid., p 13.
[98] American Psychiatric Association. Diagnostic and statistical manual of mental disorders (5th ed.). (Arlington, VA: American Psychiatric Publishing) 2013.
[99] OED Online. January 2018. Oxford University Press.
   http://www.oed.com/view/Entry/180009?redirectedFrom=simulation
   (accessed March 16, 2018)
[100] Nietzsche, F. Beyond Good and Evil, tr. Judith Norman, Cambridge University Press, p. 15, 2002.
[101] Chalmers, D. 'Facing Up to the Problem of Consciousness Journal of Consciousness Studies, 2 (3): 200–219, 1995.
[102] Plato. Complete Works Cooper & Hutchinson (eds.) (Hackett Publishing Company); 158b-d, p. 176, 1997.
[103] Zhuangzi. The Essential Writings, tr. Brook Ziporyn (Hackett Publishing Company), pp. 85,86,89, 2009.
[104] Nikhilānanda, S. Māṇḍūkyopaniṣad with Gauḍapāda's Kārikā and Śankara's Commentary (Advaita Ashrama), pp. 90, 97-99, 1987.
[105] Descartes, R. The Philosophical Writings of Descartes tr. J. Cottingham, R. Stoothoff, & D. Murdoch Cambridge: Cambridge University Press, Volume II, p. 13, 1985.
[106] Monier-Williams, M. A Sanskrit English Dictionary, Oxford, The

Clarendon Press, p. 733, 1899.
online version:
http://www.sanskrit-lexicon.uni-
koeln.de/scans/MWScan/2014/web/webtc/indexcaller.php
[107] Quoted in: Siderits, M. Buddhism as Philosophy An Introduction (Hackett Publishing Company), p. 168, 2007.
[108] Conze, E. 'Vajracchedikā Prajñāpāramitā' Serie Orientale Roma XIII (Roma Is. M. E. O.), p. 92, 1974.
[109] Ziewe, U. Vistas of Infinity How to Enjoy Life when you are Dead (Lulu), p. 17, 2016.
[110] Tholey, P. 'Consciousness and Abilities of Dream Characters Observed during Lucid Dreaming' Perceptual and Motor Skills Vol 68, Issue 2, pp. 567 - 578, 1989.
[111] McNamara, P., Dietrich-Egensteiner, L., & Teed, B. Mutual dreaming. Dreaming, 27(2), 87-101, 2017.
[112] Jung, C. Two Essays on Analytical Psychology, Bollingen, 1966.
[113] Jung, C. Archetypes and the Collective Unconscious, Bollingen, 1969.
[114] webpage: http://www.drogmi.org/the-sakya-tradition/lam-dre/the-history-of-mahasiddha-virupa
[115] http://henry.pha.jhu.edu/aspect.html

Chapter 10 Digital Pantheism

[116] Groeblacher, S. et al. An experimental test of non-local realism, 2007, arXiv:0704.2529.
[117] Oizumi, M; Albantakis, L; Tononi, G. "From the Phenomenology to the Mechanisms of Consciousness: Integrated Information Theory 3.0". PLoS Comput Biol, 10(5): e1003588, 2014.
[118] Tegmark, M. The Mathematical Universe, 2007, arXiv:0704.0646.

Chapter 11 Ouroboric Simulist Cosmology

[119] Schmidhuber, J., A Computer Scientist's View of Life, the Universe, and Everything, in Freksa, C. Foundations of Computer Science, Potential Theory Cognition, pp. 201-208, 1995, http://www.idsia.ch/~juergen/everything/

[120] Deutsch, David, The Structure of the Multiverse, 2001, arXiv:quant-ph/0104033.
[121] https://en.wikipedia.org/wiki/Path_integral_formulation
[122] http://en.wikipedia.org/wiki/Philosophical_zombie
[123] Lloyd, S. et al., "The quantum mechanics of time travel through post-selected teleportation", 2010, arXiv:1007.2615.
[124] Linde, A., How many universes are in the multiverse?, 2010, arXiv:0910.1589.

Chapter 12 "When the map becomes the territory"

[125] Klee Irwin, Code Theoretic Axiom, 2017. http://www.quantumgravityresearch.org/wp-content/uploads/2017/03/The-Code-Theoretic-Axiom-02.17.17-final-KI.pdf
[126] S. Kaufman. Unified Reality Theory: The Evolution of Existence Into Experience, Destiny Toad Press, 2002.
[127] Wai H.Tsang, The Fractal Brain Theory, Lulu press, 2017.
[128] Cosmin Visan, The Self Referential Aspect of Consciousness, Journal of Consciousness Exploration & Research 8 (11):864-880, 2017.
[129] C.M.Langan Cosmos and History: The Journal of Natural and Social Philosophy, vol. 13, no. 2, pp.313-330, 2017.
[130] J.L.Borges, 123: Narraciones (Letras Hispánicas) Tapa blanda, 2005.

Chapter 13 Reality, the Mystical Self-Referential Descent into Imagination

[131] Ben Goertzel. Creating Internet Intelligence: Wild Computing, Distributed Digital Consciousness, and the Emerging Global Brain, IFSR International Series on Systems Science and Engineering, Vol. 18, Kluwer Academic/Plenum Publishers, 2002.
[132] Tolkien, J.J.R., The Silmarillion, Del Rey, 2000 (orig. 1977).

Chapter 14 Truth-hidden-in-plain-sight: The clues for Reality as a Simulation

[133] http://www.biblewheel.com//GR/GR_37.php

[134] Tuynman, A. Technovedanta 2.0, Transcendental Metaphysics of Pancomputational Panpsychism, Lulu Press, 2017.
[135] Ervin Laszlo. Science and the Akashic Field, Inner Traditions, 2004.